WHAT IS INTELLIGENCE?

The Darwin College Lectures

What is Intelligence?

EDITED BY JEAN KHALFA

CAMBRIDGE
UNIVERSITY PRESS

Published by the Press Syndicate of the University of Cambridge
The Pitt Building, Trumpington Street, Cambridge CB2 1RP
40 West 20th Street, New York, NY 10011–4211, USA
10 Stamford Road, Oakleigh, Melbourne 3166, Australia

First published 1994
First paperback edition 1996

Printed in Great Britain at the University Press, Cambridge

A catalogue record for this book is available from the British Library

Library of Congress cataloguing in publication data applied for

ISBN 0 521 43307 X hardback
ISBN 0 521 56685 1 paperback

CONTENTS

Introduction
What is intelligence?

JEAN KHALFA

Though everyone agrees that there could be no science without intelligence, the very existence of intelligence has often been seen and used as an argument against the modern scientific explanation of the world. Is not intelligence that makes living beings act in a non-mechanical and often unpredictable way? It is not that their behaviour is erratic, but, rather, that what determines it is of the order of reasons, not of causes. What modern science does, however, is precisely not to look for the reasons stars or atoms may have for moving the way they do, but to pinpoint the causes that determine their movements. The dominant model of the universe before the seventeenth century was largely inspired by observation of the living world: each thing seemed to have its own purpose or finality – in the same way as, for instance, the very coexistence of the parts of an eye could only be understood by reference to the function of the organ as a whole: seeing – and, in turn, all the particular finalities found their reason in the ultimate purpose of a supreme intelligence. Intelligence was, then, a general means of explanation. However, as soon as the dominant model of the universe became the mechanical interaction of non-living things, intelligence obviously had to be explained away.

There were two ways of doing this. The 'dualists' postulated that intelligence was a faculty exclusive to beings ruled by an immaterial substance, a soul, of which one of the clearer manifestations in the world was the command of speech – or rather language, for parrots speak, but, as Descartes said, what they say is not *à propos* (that is, at the same time appropriate to the particular circumstances and denoting something), except by accident. It is just an acquired physical reaction to modifications in their environment that may have nothing to do with the meaning of what they 'say'. Not that in this case there could be no regular relation between the environment and the utterance: one could train a parrot to say 'What a lovely day!' whenever the sun shone. But what is wrong here is precisely that the link is immediate: the environment triggers the utterance irrespective of its content (the parrot could be re-trained to produce the same utterance whenever it rained). By contrast, appropriateness is not a causal relation, but a referential one, in which what directly refers to the environment is the content of the utterance, not the utterance itself. Now, contents of language are necessarily more than inert images, even if linguistic reference is often built on images. It is true that abstract words derive from metaphors (the word 'metaphor' for instance, is built on the image of the transportation of an object from one place to another in space, though nothing of the sort happens in a metaphor), but this is precisely what makes linguistic references something more than images (or than what images are usually taken to be, mere reflections): they are *acts* of isolation or abstraction of certain features or states of an 'outside', whether to refer to the world, to 'inner' psychological states, or to pure abstractions. No wonder then that from the consideration of the nature of language one should be led to postulate the existence of a separate 'inside', a soul, as the *agent* of the intelligent activities it requires.

Tool-making has been generally considered as another distinctive property of intelligent beings, in so far as it implies the ability to endow a material object with a mode of existence determined by an aim largely unrelated to the nature and origin of the object (as when one breaks off a branch to use it as a stick). But the use of speech can itself be seen as an act of tool-making in which some material objects, sounds, are so to speak detached from their own nature as instinctual reactions (as in cries), to be attributed meanings totally unrelated to their biological and physical nature, that is conventional ones. This is clearly shown not only by the diversity of languages invented by beings endowed with the same speech organs, but also, again, by the fundamental capacity of language to express abstractions, that is things that could not possibly have any onomatopoeic relationships with sounds.

Thus, if the dualist point of view is right, that is if only those beings whose behaviour implies the actions of an irreducible autonomous (i.e. self-regulatory) soul can be called intelligent, then intelligence must be the subject not of a deterministic science of nature (physics) but of a science of non-physical or non-solely-physical beings: men, angels and gods (metaphysics). While intelligence had formerly been the dominant principle of explanation, exhibited in various degrees in the amazing properties of even the smallest 'creatures', the gen-eralisation of the mathematical approach to science made it the privil-ege of a single species in the natural world: man. Again Descartes pointed out that the stupidest human being was in a different league from the seemingly most intelligent animal.

Such a position can also be defended from a non-metaphysical point of view. Thus Daniel Dennett, one of the contributors to the pre-sent volume, has stressed the importance and the systematicity of the different 'intentional stances' we adopt, for instance towards men-tally handicapped persons or very young children, whose behaviour we often perceive as being more determined by causes than by intentions.

This difference of attitudes not only shapes the way we interact as individuals; it is also fundamental to the institutions we have developed to regulate our interactions. For instance, most legal and, as a consequence, penal systems now in place throughout the world, are based on the notion of *responsibility*. Can one consider a non-intelligent being as responsible for the damage it may cause? Being responsible means, at the very least, being able first to foresee the consequences one's own course of action could have, and, secondly to perceive the relevance of moral, social, or legal principles to those consequences, whether or not one then decides to comply with them.

There is no doubt that both abilities are essential to the notion of intelligence. So, whether the assertion of an irreducibility of intelligence is based on metaphysical considerations about real differences between a physical and a metaphysical world, or on the analysis of the assumptions that regulate our interactions with one another (from 'folk psychology' to legal theory), from this point of view, intelligence is not a question of degree or quantity, it is an absolute or qualitative difference.

Nevertheless, there has always been an opposite approach, first criticising, and then, as science grew in strength, aiming to eliminate this last residue of a 'prescientific' conception of the world in which men, not being aware of the causes that determine their actions, have considered themselves as ultimate causes, determined only by their own ends or finalities (this is one of the origins of the usually derogatory characterisation of such an attitude as 'finalist'). As Spinoza said, they 'seemed to have conceived man in nature as a kingdom within a kingdom'. Great hopes were placed, from the beginning of the twentieth century, in the idea that in the same way as the physical complexity and adaptation of living beings could be explained by the gradual and completely mechanical process of the selection of features that simply allowed an organism to survive within a competitive natural context, the complexity of intelligent behaviour could be reduced to the selection of sets of reactions mechanically elicited by the expectation of some reward experienced in the past in similar circumstances. This had two important implications. First, the link between the behavioural reaction and what triggered it could be purely arbitrary: it did not consist in any kind of understanding. And, second, if intelligent behaviour could be reduced to combinations of simple mechanical reactions, such behaviour was not specific to one species but could be found, to varying degrees, in almost all species. This is what the most sophisticated exponent of 'behaviourism', B. F. Skinner, tried to show in his famous experiments: a rat could be trained to press a lever to obtain some food whenever a light was on (or off, according to complex schedules). As this is obviously not an innate pattern of behaviour, if one were to witness it without being aware of the past training history, one could not but think that it required some sort of intelligence. The behaviourist's point was thus that if we could produce such a result by the mere conditioning of a physical reaction to any 'stimulus' with which it had usually been associated in the past with a beneficial effect, there is no reason to assume that other types of apparently intelligent behaviour could not be explained in the same way, removing the need to refer to metaphysical entities, or to use non-causal, finalist modes of explanation.

The problem is that these experiments were too successful. They showed that animals could be very intelligent indeed, too intelligent in fact, for the theory. For if the rat's paws were tied, it used its nose; if a lever were shortened, it did not press in the air but moved forward. Which means that what had been conditioned in the experiment was not simply a specific muscular reaction to a particular stimulus, but some understanding of a particular causal state of the world. The difference is that once the link was understood, a large class of possible actions (instead of a single fixed reaction) could be executed, of which the only common characteristic was intentional: they would produce the desired effect. The action in the contrived environment of the box was only fixed or regular because pressing a lever with one's paw or hand rather than nose is the most natural thing to do in such circumstances, whether one is a rat or a man (but not when one is a seal). Therefore, though associative learning theory had produced some astonishing results, and also demonstrated the ability of human intelligence to invent unexpected methods of enquiry, it could not meet its eliminative goal: intelligence remained a problem that could not be solved by a simple mechanistic approach. And if much has been done lately in the cognitive sciences to tackle the problem of intelligence, there is still no agreement as to whether solving it will be a process of reduction to the biological and physical sciences, or whether we should review the way we think knowledge should be organised and unified.

The emergence of intelligence is an evolutionary problem: why is instinct suddenly insufficient at a particular stage in the history of adaptation? Are there necessary limits to instinctual solutions when the living world has reached a particular degree of complexity? And anyway, is the difference between instinct and intelligence such a clear one? After all, one of the effects of the 'cognitive revolution' is that in all corners of the life sciences we now find intelligent processes at work. Finally, are there degrees of intelligence, as of other natural faculties? It is because of questions such as these that this book is organised along an evolutionary plan, and starts paradoxically by examining this apparently most passive of life's functions, perception, at the level of the brain, to finish with a study of language. However, such a plan should not be taken to suggest that the evolution of intelligence is in itself linear and progressive, reaching its pinnacle in the human mind. Most of the papers have at least one conclusion in common: there are varieties of intelligence and they cannot be easily compared, let alone rated on a common scale.

In the first chapter, Richard Gregory deals with visual perception, and shows that a great deal of intelligence is required by what seems *prima facie* a receptive state. In fact what this seminal paper tells us about perception leads to an understanding of general features of intelligence, and in particular of the fact that it is a property of *processes*, not of predetermined beings. Thus, visual perception can be seen as the result of the work of particular forms of intelligence: on the one hand, the sedimentation of evolution in the physiological forms of sense organs (the store of solutions or the 'knowledge' that life has evolved, what Gregory calls its 'potential intelligence'), and, on the other, the actual use of decision-making mechanisms when the amount of sense data available either under-determines or overwhelms the assumptions embodied in the physiology of the organs, and leads to perceptive ambiguities in the identification of an object or a phenomenon (this inventive type of intelligence he calls 'kinetic intelligence'). These considerations set out the problem clearly, both in theoretical and in practical terms: the main theoretical question, which runs throughout the book, is that of the respective importance first of knowledge and secondly of the means and ways of accessing and using it. An important practical question is the validity of attempts at measuring degrees of intelligence on a single scale, such as IQ tests. If they are so unsatisfactory, it is precisely because they reduce intelligence to only one of these aspects, turning it into an empty formal ability.

In chapter 2, Nicholas Mackintosh systematically compares attempts at explaining animal behaviour on the basis of associative learning with intentional approaches. While in most tasks involving the perception some animals have of abstract relations the first approach fails, the intentional approach appears unnecessary in unexpected and important cases, such as communication, where most of what seems to require the attribution of beliefs by some animals to other ones can in fact be reinterpreted from the point of view of an associative theory of learning, as techniques for manipulating the other's behaviour. Here again it seems difficult to point to a unique dimension of intelligence, different species being better or worse than others at different tasks, and sometimes in ways that contradict our expectation of what the hierarchy of intelligence among them should be.

At the next level, that of the emergence of intelligence in humans, George Butterworth re-examines the best-known theory of the development of intelligence in children, that of Jean Piaget, and shows that in many respects, the idea of a regular development through necessary stages, according to the degree of physical interaction the child has with the world, proves inadequate: several fundamental concepts (such as that of 'permanence', that is the assumption that objects that have been removed from perception continue to exist) are in fact present at a very early stage and are best explained by an examination of the structure of perception. Also, children who have severe motor handicaps from birth can develop intellectually in roughly the same way as children who can interact normally with their physical environment. Again, processes that we would want to call intelligent are at work very early on in the structure of perception itself. The driving force of the development of intelligence seems to lie more in the ability to translate a perceptual understanding from one sense to another, or to transfer it to a different type of activity, that is in the ability to generalise solutions that are already present. By contrast, stupidity is a form of absolute specialisation.

Having acquired a better understanding of the diversity of intelligent processes in nature, it is only natural to consider the possibility of an artificial intelligence (AI). In chapter 4, Roger Schank and Lawrence Birnbaum forcefully assert that the problem is similar to the one faced by educators; intelligence is a question of learning, of acquiring a memory or knowledge-base of sufficient size, and of developing the retrieval mechanisms necessary to use it. Since 'AI is in the size', creating such an intelligence is largely a progressive or cumulative process. The strength of the argument is that since the alleged theoretical problems of AI are in fact pedagogical ones, the solutions AI researchers devise can also offer original ways round pedagogical problems. Conversely, most of the objections raised against it come from conceptions of intelligence and knowledge that ultimately deny the possibility of true learning or teaching.

Schank and Birnbaum's claim of a continuity of intelligence between all species, natural and artificial, is obviously quite controversial. To measure its implications and also to understand those who disagree with such a claim, it was necessary at this point to examine specific features of human intelligence. The second group of chapters look at two intelligent activities (mathematical science and art) where invention is crucial, before moving on to language.

In chapter 5, Roger Penrose shows that the process of inventing a proof is often a particular type of visualisation much more than a 'blind' deduction. A good comparison might be to see it as an orienteering race, as opposed to a single line road race, except that in his thought-orienteering, the mathematician does not have a predefined map of all the elements he will need, nor of all the potential paths to the solution. If he is to progress, he has to find a way to reduce the world of all possible paths. By demonstrating the impossibility of the mathematical mind having *a priori* 'frames', Penrose wants to show that there is in mathematical intelligence a non-deductive element that eludes any computational description. Thus, paradoxically, the mathematician's mental space is itself more geographical than geometrical: susceptible of exploration but not of demonstration (except for the demonstration of this impossibility). And again, such a discussion is not without practical pedagogical consequences, in particular as regards the specific problems of teaching mathematics. As structuring a space and orienteering are among the first things we do, even before being able to move (see what Butterworth says about the infant's ability to follow its mother's gaze), it could be inferred from Penrose's argument that teaching methods which insist on the early acquisition of strictly deductive procedures, as opposed to the traditional ones which favoured applied or manipulative geometry, are deprived of a natural way of familiarising the child with mathematical reasoning.

Penrose's argument against the ability of an artificial intelligence to emulate fully the capacities of human intelligence also suggests that knowledge and intelligence do have important differences. For if, on the one hand, there cannot be any *a priori* plan of the best ways to solve all problems, and if, on the other, an infinity of consequences can be logically derived from any particular course of action (most of them irrelevant, but without there being any *a priori* rule to eliminate them), then a finite intelligence cannot base its decisions on the systematic and exhaustive consideration of the consequences they may have. In fact, supposing that an infinite mind existed, able to examine all future moves on the infinite chessboard of the universe, we would still be quite reluctant to call it intelligent, precisely because it would think in a 'mechanical' way. Would this mean that God cannot be intelligent? That would seriously jeopardise the possibility of His existence, for a perfect being must have, by definition, all conceivable faculties. A way around this consequence would be to say that the only humanly conceivable reason for man's freedom (supposing that this is a fact irreducible to the deterministic laws of nature) is precisely that if the world were devoid of uncertainty, God would have no opportunity to apply His intelligence in trying to understand it (surely, a great deal of His intelligence would be needed to understand man's behaviour), and it would be absurd for a perfect being to be endowed with a useless faculty. Hence man's freedom would be a proof of the existence of an intelligent God (rather than the reverse). Another, more likely argument, would be to say that in fact intelligence is not a positive faculty, but simply an ability to guess successfully, a way of coping with limitations either in knowledge or in the ability to access the relevant part of it in time. Thus, by definition, God would not *need* to be intelligent. Which is not to say that if God exists He is necessarily stupid, for, as we have seen, the notion of stupidity is also linked to that of a limitation of knowledge, but more in a horizontal way, as a lack of breadth.

While knowledge tends towards unity, it seems that intelligence is a process of production and diversification of forms or patterns of understanding, which can shape empirical data to create knowledge, but have much wider cultural uses. An ethnologist is best placed to illustrate the diversity of human intellectual achievements, in particular when dealing with music, where such diversity can clearly be experienced and measured. In chapter 6, Simha Arom shows how it is possible to provide people living in societies where transmission of musical knowledge is oral, and based largely on example and imitation, with traditional instruments that have been cleverly computerised so as to give the musicians the capacity to examine, and work on, the very complex mental structures they were previously simply inhabiting. It thus appears, for instance, that Central African xylophonists build and tune their instruments in an ambiguous way (between two scale systems), not through lack of precision, but so as to be able to accompany on one instrument songs sung in different modes. This means that these cultures have developed a meta-knowledge of their own music, though there is no explicit theory of it. It is difficult to convey in a book the experience of understanding a foreign musical idiom through its live deconstruction and reconstruction, but what is clear here, is that there is not *one* artistic intelligence: the range of possibilities within which the mind works is wide enough for it to generate and apprehend extremely complex and varied forms. Aesthetic pleasure is perhaps precisely the pleasure of experiencing one's own ability to invent such complex forms, either during the creation/performance of a work or through its reception. Looking at a painting or listening to a piece of music are not just seeing or hearing, they are activities in which one *creates* forms or paths of perception. Works of art might differ in aesthetic value in accordance with their ability to elicit such reactions.

Ultimately, an examination of language is crucial here, first because of its essential role in the development of human intelligence, and secondly because of the intelligent procedures it embodies. What Daniel Dennett shows in chapter 7 is that language is as essential to the evolution of mind as competition and selection are to the evolution of life. He builds up a metaphorical tower representing the possible degrees in the hierarchy of control-systems that evolution is able to develop in the brains of various species. This might be called *Dennett's Anti-Babelian Tower*: language is almost at the top, not only as what allows the creation of an inner milieu, an ideal mental space within which we can experiment with the world without danger, but also as what gives us access to the mental space and world experience of countless other intelligences, with whom we have no genetical kinship, through space and time. While individual animals directly exemplify their genetic inheritance, language allows the identity of any individual human to be shaped also by the history of all the other individuals of the species, past and present. There is no doubt that seen in that perspective, language is the essential criterion of human specificity.

But is the use of language itself an intelligent process, or is it simply a mechanical activity of coding and decoding pre-existing messages? In the final chapter, Dan Sperber shifts away from the traditional emphasis on language production and structure, to concentrate on language understanding. He shows that even the simplest acts of communication convey complex hierarchies of intentions that any speaker must be able to infer if he is to understand other speakers' utterances and respond to them. Thus, as the ability to assign intentions to others must be prior to verbalisation (a recurrent theme throughout this book), the explanation of language must ultimately be grounded in the psychology of intelligence. Now, how this ability develops in children, and maybe in some animals, and whether machines can be endowed with it, will probably be at the centre of discussions of intelligence for some time to come.

Over the years, the books of the Darwin College Lectures have come to form a particular sort of encyclopedia, similar in spirit to those of the Enlightenment, i.e. centred on problems rather than organised according to the established categories of knowledge. Such books aim more to exhibit the complexity of a problem and to present the most advanced contemporary approaches to it, than to provide a final solution – a restraint particularly necessary in the case of a notion like intelligence which has only recently become a specific object of scientific enquiry. However one thing is certain: on reaching the final chapter of the book, the reader will realise how much more intelligent he is than he ever thought (though this may not make him more intelligent than he actually was all along).

1

Seeing intelligence

RICHARD GREGORY

What is intelligence? Intelligence is hard to define and descriptions are generally beset with paradoxes. Thus intelligence is attributed to those who have to think because they do not know a lot, and to those who know a lot and so do not have to think.

PARADOX OF INTELLIGENCE

People are said to be more intelligent when they work out an answer than when they already know it (which is a kind of a cheat), yet we also say that scholars who know a lot are more intelligent than ignorant people. This paradox may be resolved by pointing out that becoming a scholar requires intelligence – for learning is difficult. But we do speak of knowledge itself as a kind of intelligence. When Macbeth asks 'Say from whence you owe this strange intelligence?', he is asking for the source of knowledge or of information. This has the same meaning as 'military intelligence', which does not imply that military people are especially bright. On the other hand, if we say that Einstein was intelligent we refer to what he invented or discovered, rather than to what he learned at school or in later life. It is because what he said was not already known that we regard Einstein – and Newton, Faraday, Darwin and other great innovators – as exceptionally intelligent. This is not the 'intelligence' of existing knowledge. It is the 'intelligence' of discovering or creating new knowledge.

So we find two very different kinds of intelligence: intelligence of stored knowledge and intelligence of processing, for problem solving. I suggest we may usefully call these potential intelligence and kinetic intelligence. Although distinct as concepts, they are found together intermixed, for some knowledge is needed for solving problems and some initiative is needed to apply knowledge appropriately.

The terms potential intelligence and kinetic intelligence are meant to suggest analogy to potential energy and kinetic energy. Let's spell this out. Consider lifting a weight. We are using kinetic energy from our muscles to increase the potential energy of the weight – which may then be used for some purpose, such as running a clock. In a murder story the criminal carries a heavy stone up to the ledge, giving it potential energy; then the sinister murderer needs only a small shot of kinetic energy to release this stored potential energy, to crush his or her victim. To gain potential energy one may not have to lift the weight oneself, for one may get help from other sources of kinetic energy, and one can buy potential energy in many forms, such as chemical energy stored in dynamite. This also requires but small kinetic energy to release it – to use kinetic energy stored from the past, for future potential good or ill.

Returning to intelligence: we find potential intelligence, not only in brains or minds but also in books and tools. Tools contain ready-made answers to practical problems. Given a pair of scissors, you do not need to solve the problem of how to cut cloth. This problem is already solved in the design of the scissors. Scissors were developed through many generations by steps of kinetic problem-solving intelligence, to produce the potential intelligence that is built into their present design. Apart from the invented basic feature of pivoted knives, all manner of problems had to be solved, such as techniques of metallurgy and how to form their shapes and sharpen them. Each step required some kinetic intelligence to solve the problems: now these are stored as useful potential intelligence solutions built into every pair of scissors.

Scissors are tools created by human intelligence. What of biological 'tools' such as fingers and eyes? What of the almost incredible processes that imbue organisms with life, the metabolic processes that plants and animals possess and control; which for us, with all the powers of science, are extremely difficult to understand or create. No one yet fully understands photosynthesis in plants, or the processes from light in the eye to the neural impulses transmitted to the brain for seeing. Although still mysterious to us, and to science, these problems have been solved millions of years ago through the course of evolution by the processes of natural selection. So, surely, we should call natural selection intelligent. Indeed it is a very powerful kinetic intelligence that has discovered and invented answers and processes of life and mind that are largely beyond our understanding. Natural selection is the kinetic intelligence which over eons has produced the immense store of potential intelligence embodied in us. Babies do not have to invent muscles to move, or eyes and brains to see; these are already invented by the kinetic intelligence of evolution, and inherited as rich potential intelligence built up over countless generations. Considered like this, almost everything we do we owe to the immense potential intelligence we inherit from the hundreds of millions of years of kinetic invention of natural selection from the first life forms on Earth. So our brains have relatively little to do!

It might be objected that natural selection does not look intelligent, so we shouldn't call it intelligent. Random experiments, surviving threats of predators and nature to have progeny carrying successful genes, look very different from the inspired creation of human geniuses. Yet do we know that processes of brain intelligence are essentially different? It has been suggested that ideas survive or die by quite similar selection, according to their success. In any case we should not expect processes to look intelligent – intelligence is the result of processes. These should not look intelligent any more than the processes and parts of a sewing machine look like stitches; or the individual processes and parts of an engine have power. How this is done is not obvious if we don't know much about sewing machines, engines – or brains. It took the cognitive mutation of Darwin to see that species are created by selections of the fittest mutations. However it appears, natural selection is supremely intelligent if we define intelligence in terms of creating answers to hard problems. It is no objection to say it does not look intelligent; science is full of non-intuitive theories. Indeed this shows science's power to change minds – with science's stored potential and active kinetic intelligence.

Let us consider some further possible objections to calling natural selection 'intelligent'. Should it be denied intelligence because it does not see where it is going – as evolution is blind? I don't see why. Neither Darwin nor Einstein knew where their theories would lead. Intention is not necessarily involved in discovery or invention, though some are guided by intentions and predictions of what may be found. Applied science often has set intentions, though with surprising outcomes; but theoretical science follows its own paths and (in spite of pious promises to funding sponsors) is seldom precisely directed to predicted goals. The best experiments yield surprising results. So its 'blindness' should not preclude natural selection being called intelligent.

It might be objected that, although natural selection solves immensely hard practical problems, it does not produce theories or explanations. Thus the leaves of trees carry out photosynthesis, but they cannot say how or why. They are silent guardians of secrets their ancestors have discovered. But this is true also of scissors and sewing machines; they embody generations of human kinetic intelligence yet they cannot say how they work or what they do. And very few human intelligences produce original theories! It might also be objected that organisms or brains can be intelligent because they are things, while the processes of evolution can hardly be called intelligent because they do not exist, as objects. It is not the brain as an object that is intelligent, but the processes it carries out that produce intelligent solutions. It is processes that matter in the matter of the brain. Lastly, what of intentions in consciousness? Well, it is a common experience for an answer to pop up unexpectedly, without consciousness. And we do not want to include consciousness, awareness, as a necessary criterion for intelligence lest it precludes by definition intelligence in AI machines, including seeing machines.

The analogy of kinetic and potential energy to potential and kinetic intelligence seems to work rather well and might be useful. But, is the analogy to potential intelligence completely correct? When the weight is released or the dynamite exploded, potential energy is lost. Is knowledge lost when used? Is the potential intelligence of scissors lost when they cut cloth? This is not generally so, though scissors do lose potential intelligence as they wear out. Military intelligence loses its point if it becomes generally known; for this intelligence must be kept secret from the enemy to be useful. On the other hand science gets its strength from sharing knowledge. It is only secret potential intelligence that is eroded through sharing. Here the analogy breaks down for the potential intelligence of knowledge is not lost through use except where there is competition. Co-operative sharing of knowledge minimises the need for kinetic intelligence. This is the extraordinary power of science which separates us from our biological origins.

TESTING INTELLIGENCE

Testing individual intelligences in children or adults must be extremely difficult when we don't know the relative contributions of potential and kinetic intelligence. That is, when we don't know how much the answer is given from prior knowledge, and how much it is worked out at the time. It seems likely that the brain's store of knowledge (including rules or algorithms for problem-solving) is so vast it virtually swamps differences in active processing kinetic intelligence. If so, this must make measuring intelligence as conceived by psychologists almost impossible, for differences of stored knowledge (including of skills) will dominate, and so make differences of kinetic intelligence (IQ) second order.

As Sir Peter Medawar and others have pointed out, the IQ notion of intelligence as a single dimension along which to judge children and adults, is among the most personally and socially damaging notions of this century. There are lots of kinds of people – musicians, cooks, craftsmen, scholars, bankers, painters, jokers. It is invidious to place them on a single dimension. Fortunately there are lots of kinds of intelligences.

Can we define creative intelligence? There must be some novelty, or there would only be potential intelligence from the past. Precise copies are not newly intelligent: a xerox machine does not generate creative intelligence, however brilliant the documents it is copying. We may say that creative intelligence depends on some randomness, to avoid mere copying, and to escape the tyranny of rules. Randomly generated possibilities seem essential for novelty; but there must be limits to the range of generated randomness, or the chances of discovering appropriate answers would be vanishingly small. Apart from poetry, it would be pointless to set out to look for green cheese on the moon! Evidently there is a combination of restraints from past wisdom and essentially random search for new possibilities, for intelligent solutions to be appropriate and novel.

We might define kinetic intelligence as: 'generation of appropriate novelty'. What is stupidity? Lack of appropriate novelty, of course. But this can take two forms: lack of novelty is plain dull, but inappropriate ideas can be exciting, and may even lead to new worlds in science and art. These, though unconscious and undirected, may come to look supremely intelligent as they become appropriately novel.

THE (USUALLY) INTELLIGENT EYE

One of my books is called *The Intelligent Eye*. How is seeing intelligent? Do we find potential and kinetic intelligence in the procedures of vision? Answers depend on how vision itself is seen – on what kind of theory of perception one accepts. If perceptions were no more than responses to stimuli, there would only be the inherited and learned potential intelligence of appropriate reflexes. If perception is seen as involving active (though unconscious) decision making, then we allow perception some kinetic intelligence. What is the evidence? Essentially that, beyond tropisms of very simple creatures, recognition of objects from limited data of the senses requires intelligence. Perceptions are much richer than the available sensory data – and have to deal with new objects and situations – so there does seem to be a lot of intelligent creation of novelty going on, every day, all the time, whenever we open our eyes and see. But is this always appropriate novelty? No – for perceptions are quite often incorrect and suffer intriguing suggestive phenomena of illusions. These can tell us a lot about the occasional unintelligence of the intelligent eye.

ILLUSIONS

The most suggestive illusions are ambiguities of perception. The same retinal image may be seen as a variety of objects, and perception may change spontaneously from one to another. Where there are plenty of redundant data ambiguities disappear. This indicates creative decision taking for creation of objects on the limited data of sensory inputs. When perception gets it wrong it is inappropriately novel.

There are many kinds of illusions. Some are due to physical causes, such as a bent-stick-in-water, or a rainbow; others to physiological errors of signalling in the nervous system; others again to misreading the neural signals. Ambiguities – such as those below – of the reversing Necker cube, or Jastrow's Duck-rabbit are good examples of this, the same signals producing different object-perceptions as they are read differently. They produce alternative hypotheses.

Let us now try to classify visual illusions. This might be done from their appearances; but it is far more interesting to attempt a theoretical classification.

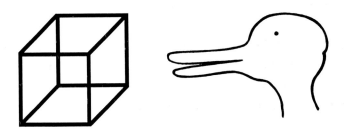

Figure 1 Left, Necker cube; right, Duck/rabbit

From the three different kinds of processes of vision, from physics through physiology to cognition, we derive three basic kinds of illusions: PHYSICAL – PHYSIOLOGICAL – COGNITIVE. It turns out that each of these can create: AMBIGUITIES – DISTORTIONS – PARADOXES – FICTIONS.

I shall now suggest examples of how some of the well-known visual illusions may be classified in these terms. Not everyone will agree, and no doubt with increasing understanding many of these phenomena of illusion will change categories. This is typical of classifying phenomena of the natural sciences, and should be expected for this unnatural science of illusions.

I **Physical illusions**	II **Physiological illusions**	III **Cognitive illusions**
AMBIGUITY Mist Shadows	AMBIGUITY Size-distance, for a single eye Ames Demonstrations	AMBIGUITY Necker cube Duck-rabbit Vase faces Hollow mask

DISTORTION
Of space: Stick-in-water
  Mirages
Of velocity: Stroboscope
	DISTORTION
	Adaptations to length,
	  tilt or curvature
	The 'Café Wall'
		DISTORTION
		Muller-Lyer 'arrows'
		Ponzo 'railway lines'
		Poggendorff displacement

PARADOX Mirrors (Seeing yourself through the mirror, yet knowing you are in front of it)	PARADOX Rotating spiral after-effect (Expands yet gets no larger – paradox when visual channels disagree)	PARADOX Impossible objects Escher's pictures

FICTION
Rainbows
Moiré patterns
Holograms
	FICTION
	After images, autokinetic
	  effect
	Migraine patterns
		FICTION
		Kanizsa triangle
		Filling-in of the blind spot
		  and scotomas

It might be objected of the physical origin of illusions, that the physical states are not themselves ambiguous, distorted, paradoxical or fictional – but this is not the suggestion. Mist originates visual ambiguity though misty air is not itself ambiguous. Mirrors produce visual paradoxes though they are not themselves, as objects, paradoxical. Whether illusions are produced from 'physical' origins depends on physiological and cognitive limitations of perception. Diving birds avoid errors from the bent-stick effect of refraction by water though we are fooled. In principle we might literally see through refractive distortions, and mirror paradoxes; but this is beyond our perceptual abilities. We can only see through them conceptually, by understanding them.

It is not hard to add sub-classes under these major categories. Thus physical includes many kinds of distortion, loss, and additions to patterns of energy (stimuli) upsetting evidence of objects. The most dramatic physical creation – fiction – is holograms. Many physical illusions are described by Tolansky, and a very full account is given by Minnaert of naturally occurring and especially meteorological phenomena giving physical illusions.

Physiological illusions include neural losses and interactions, producing degenerated or distorted neural signals and sometimes spurious activity producing bizarre fictions, including those of drugs. Presumably apparitions of schizophrenia are in this category, though some cases may be due to bugs of cognition.

Figure 2 Bars. Stare at the horizontal line on the right for about ten seconds – then look at the central dot. The central vertical bars should look twisted in an opposite direction to the right hand bars. The left hand bars affect the size (spatial frequency) of the central bars. These are adaptations of specific visual channels, for orientation and spatial frequency

Cognitive illusions are the most subtle and the most controversial. How does one show that a perceptual phenomenon is cognitive rather

Figure 3 Hollow faces

than physiological? One has to show that it is due to (mis)use of assumptions or knowledge; or to inappropriate rule following, leading perception astray. Here we reach tendentious issues, of how much of perception is given 'Bottom-up' from sensory signals and how much 'Top-down' from stored knowledge. A clearly Top-down, knowledge-based illusion is the Hollow Mask. A hollow face (the inside of a mask) looks like a normal nose-sticking-out face. This is so even when it is viewed with both eyes from quite close. That the hollow mask looks like a normal face is evidently due to the low probability of a face being hollow – so this very powerful illusion reflects our knowledge of faces. We may call this effect 'shape-from-knowledge'. For the hollow mask, our knowledge of faces is inappropriate – so there is a (cognitive) illusion. When the mask is rotated, it apparently rotates in the reversed direction, as motion parallax is mis-ascribed to what is near or more distant: so one illusion generates another.

The use of very general rules, such as the gestalt principles – of proximity, common fate, closure and so on – is different, as they apply to almost all objects, but do not reflect knowledge of any particular objects. Some rules are employed in all situations; but others may be selected according to need. Presumably perceptual processing is somewhat different for different tasks: reading, walking, putting up shelves and so on. So we might think of operating rules introduced 'side-ways' according to need, like floppy discs. Illusions will be produced when the selected disc's rules (or algorithms) are inappropriate for the task. There is a shortage of special 'floppy discs', so there are many rule-generated cognitive illusions.

This concept of inappropriateness of assumptions and general operating rules generating illusions is not altogether popular with physiologists. They are not due to malfunctioning of physiology, so they do not have physiological explanations – and so do not belong to physiologists! They are seen with inappropriate 'floppy discs' of visual rules.

The 'geometrical' illusions seen in perspective drawings are probably due to mis-setting size constancy – as for flat pictures the normal compensation for retinal images shrinking with increased object distance is inappropriate. This notion is suggested and supported by the observation that depth-cues indicating greater distance are associated with illusory expansion of size. Also, the distortions disappear when the perspective is appropriate to the seen object. Evidently

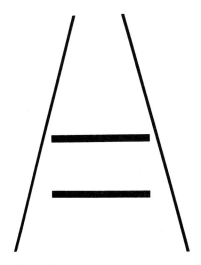

Figure 4 Ponzo

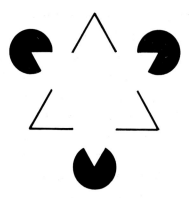

Figure 5 Kanisza triangle

there are no special 'floppy discs' for pictures – presumably because there is no behavioural feedback from pictures, which are not handled and the distortions do not produce behavioural errors. If one handled pictures, or Adelbert Ame's queer-shaped objects, so that the illusions mattered then our visual system might write special 'floppy discs' for them. But such special 'floppies' as for seeing pictures are in short supply – hence these cognitive distortion illusions.

To make the study of illusions a proper unnatural science, litmus tests are needed to place the phenomena in their appropriate categories. This, however, takes us to too technical considerations. An interesting example is whether a distortion illusion is lost when seen in correct perspective. This happens for cognitive misplaced constancy distortions but not for physiological signalling errors.

Among the most interesting illusions are cognitive fictions. We all have a large blind spot in each eye. This we can 'see' by making one of the two symbols below disappear. Try looking at the right-hand symbol with the left eye only (covering your right eye with your hand), from a distance of about ten inches. The left-hand symbol will disappear as it falls on the blind spot where the optic nerve leaves the retina for the brain.

£   •

But we don't see a great black, or blank, region at the side of vision. Is it simply ignored, or does it get actively filled-in by a (cognitive) brain process? Recently, V. S. Ramachandran and I have found that brain evidently creates to fill in small blind regions of the eyes from with the naturally occurring blind spot, and with blind regions created artificially by adaptation, it is possible actually to see what the information from the surrounding pattern. The experimental technique, using computer graphics, is to look very steadily for about ten seconds at a pattern having a blank region of the same mean brightness as the pattern. The blank region gradually disappears. Then a blank screen, of the same colour and brightness, is substituted on the computer. In this second field one sees for a few seconds, in exactly the place where the blank region was in the pattern of the first field, a rough version of the original pattern of the first field; though it has been entirely removed. Evidently this rough version of the pattern is generated by the brain – like a virtual reality – to fill the missing region, and remaining visible for a few seconds on the blank screen. So we can see cognitive fictions created to fill blindnesses. This works even for dynamic patterns of twinkle, or visual 'noise'. By making blind regions visible and discovering how and when they are filled, we might aid the eye surgeon's decisions when damage must be done to heal and help his patients to make maximal use of impaired sight. Though the intelligence of evolution is blind, the intelligence of the eye can create fictional vision to hide some of its own blindnesses.

FURTHER READING

Gardner, H., *Frames of Mind*, London: Heinemann 1984.

Gregory, R. L., *Eye and Brain*, London: Phoenix 1966. Fourth edn. 1990.

Gregory, R. L., *The Intelligent Eye*, London: Weidenfeld 1970.

Gregory, R. L., *Mind in Science: A History of Explanations of Psychology and Physics*, London: Weidenfeld and Nicolson 1981.

Gregory, R. L., Intelligence based on Knowledge – Knowledge based on Intelligence. In R. L. Gregory and Pauline K. Mastrand (eds.), *Creative Intelligences*, London: Francis Pinter 1987.

Ittleson, W. H., *The Ames Demonstrations in Perception*, Princeton: Princeton University Press 1952.

Kanizsa, G., *Organization of Vision: Essays on Gestalt Perception*, New York: Preager 1979.

Minnaert, M., *The Nature of Light and Colour in the Open Air*, London: Dover 1954.

Rolt, L. T. C., *Isambard Kingdom Brunel*, London: Longman 1970.

Sacks, Oliver, *Migraine: Evolution of a Common Disorder*, London: Faber and Faber 1971.

Tolansky, S., *Curiosities of Light Rays and Light Waves*, London: Vaneda 1964.

2

Intelligence in evolution

NICHOLAS MACKINTOSH

When we look back at the past from our present vantage point, we are often tempted to view it as a series of faltering steps leading to the present; to interpret past events in the light of our present concerns, to judge them as more or less modern or prescient; and to look for signs of progress. Thus the Whig interpretation of English history in the seventeenth and eighteenth centuries judged that history simply as progress towards the happy, prosperous, liberal England of Victoria. On a longer timescale, late nineteenth century histories of civilisation judged earlier societies and other groups of humankind as more or less savage, more or less on the road to modern western civilisation. On an even longer timescale, evolutionary theorists have often been inclined to see the course of evolution as a series of ascending forms of life leading up to the emergence of modern humans, the pinnacle of the evolutionary process.

EVOLUTION AS PROGRESS

Nowhere is this temptation more insidious than in comparative psychology, when we start talking about the evolution of human intelligence. We all believe that we are more intelligent than other animals, and most of us believe that evolution has involved a progressive increase in intelligence as we ascend the phylogenetic scale, from invertebrates to lower vertebrates, to mammals, primates, our cousins the great apes, and at last ourselves. A further source of temptation here is that we have a single word, 'intelligence' for what we are talking about, as though there were a single thing or unitary trait for that word to refer to. If intelligence is unitary, it follows that we might be able to rank-order animals by how much they had of it, just as IQ testers are inclined to rank-order people by the amount of intelligence they are supposed to possess. Most of us seem to find it perfectly reasonable to be asked whether a monkey is more intelligent than a bird. People tend to agree in the answers they give to such questions, ranking apes above monkeys, and monkeys above dogs and cats – though there is some dispute about which of those two is cleverer. Cats and dogs are seen as cleverer than horses; horses, in turn, are cleverer than cows and sheep, which are certainly more intelligent than chickens, which are probably more intelligent than fish.

One consequence of this string of beliefs is that most attempts to study animal intelligence have consisted simply of a search for the precursors of human intelligence. Even Darwin himself, in *The Descent of Man* (1871), was concerned largely to find elements of human intelligence in other animals – of human language in bird song or the vocal imitations of parrots, and of human conscience in the behaviour of a dog towards its master. Darwin at least saw that it was important to decompose human intelligence into component processes or operations. In the hands of his successors, however, the enterprise rapidly reduced to the compilation of anecdotes designed to illustrate, in a wholly unanalysed way, how clever or human-like animals were.

I want to argue that this is an almost wholly erroneous view of the evolution of intelligence; that there is no single thing called intelligence; that we certainly cannot rank-order animals by their possession of it; and even more certainly that there is no linear progression from lower to higher vertebrates, lower to higher mammals, and then onwards and upwards to monkeys, apes and humans. Intelligence is much more profitably viewed as a diverse or heterogeneous array of processes, operations and skills. Some of these are general and surprisingly widespread – being common, for example, to essentially all vertebrates; some are specialised for particular purposes and tasks and may be much less evenly distributed. But there is surprisingly little evidence to support the view that more advanced forms of intelligence are to be found only in higher animals.

SPECIALISED ABILITIES

I start with a proposition that should be reasonably uncontentious. There is no doubt that different animals have different specialisations. Other than primates, few mammals share our capacity for colour vision, but some other animals can detect wavelengths well outside our range of sensitivity. Some can detect the plane of polarisation of light; others sounds of substantially higher frequency than we can hear; and the olfactory world of most mammals is incomparably richer than ours. Of course, one is inclined to say, these are merely differences in sense organs: there is nothing mysterious in that – and nothing to do with intelligence.

There is, however, a great deal about animals that both remains mysterious and is clearly very complicated – surely intelligent in one of the senses that Richard Gregory describes in his chapter. For example, bats that live in caves fly around at high speed in pitch darkness, navigating by sonar or echolocation. But how does each bat discriminate the echoes of its own sound pulses from the hundreds or thousands of echoes generated by its fellow bats, all flying around doing exactly the same thing? There is equally much that remains mysterious in the ability of migratory birds or homing pigeons to navigate across hundreds or thousands of miles of unknown territory; and some parts that we do understand of this process are actually remarkably complicated. For instance, birds (and honey bees come to that), use the sun's azimuth (or horizon) position as a compass – a system that not only requires an accurate internal clock, since the position of the sun varies with the time of day, but can also be shown to involve the ability to extrapolate from the sun's present rate of movement to its expected azimuth position several hours later, so that they can calculate what the angle between the sun and their goal will be at that time. That this is an intelligent system is paradoxically shown, as Gregory has argued elsewhere, by the errors the bird or bee makes. The honey bee can only extrapolate from the current rate of change of the sun's azimuth position, but since that changes during the course of the day, they make systematic errors if required to predict too far ahead.

It is not only in spatial perception that we are outclassed by some animals; others outclass us in spatial memory. Clark's nutcracker, a bird living in the Rocky Mountains with a brain weighing less than ten grams, feeds off pine seeds – a commodity in scarce supply for much of the year, but when it comes, in autumn, available in glut. The nutcracker stores them for its food supply throughout the winter, but in order to do so has to hide them well, and not all in one place for otherwise someone else will find and eat them. Observation in the wild suggests that a single nutcracker hides some 30,000 seeds, more than half of which it is able to find over the next six months. How? By remembering where it hid them. Laboratory experiments reveal a spatial memory for hidden food far superior to that of the human experimenter, who must laboriously record where the bird has hidden the seeds. And other laboratory tests of spatial memory reveal, perhaps unsurprisingly, that nutcrackers outscore other non-food-storing birds, like scrub jays or pigeons.

GENERAL ABILITIES

These may all seem like examples of perception, not of intelligence, even if tests of spatial orientation and memory appear in human IQ tests. Intelligence, many people will want to say, is about understanding the world not just perceiving it; predicting the future; adjusting to changing circumstances; solving problems; drawing inferences; reasoning. But this brings me to the common general processes. For contrary to much received opinion, it is reasonable to argue that the major processes by which animals achieve many of these goals are both widespread, that is to say common to most if not all vertebrates, and relatively simple. I refer here to basic associative learning or conditioning – Pavlov's dog salivating at the sound of the bell that signals its dinner, or the rat pressing a lever in a Skinner box to earn its pellet of food.

The common perception of conditioning experiments is that they are artificial, irrelevant and boring. Of course they are artifical in the sense that most laboratory experiments are artificial abstractions from the real world; and so, on the whole, they need to be if they are to find anything out. Irrelevant? This seems hard to sustain. Conditioning provides the means by which animals learn to return to places where they have found food before, learn what food is good for them and what bad, and acquire the skills needed to deal with that food (for example, to capture prey or open up inaccessible food). It also enables animals to learn where predators may be lurking, and how to take appropriate action at the first signs of danger rather than wait for the predator to pounce. A process that achieves all this is hardly irrelevant to most animals' daily life. And the notion that conditioning is boring stems from an outmoded conception of the nature of associative learning – that it is simply a matter of the stamping in of new reflexes, blind, mechanical and mindless. This seems a serious misconception. Consider, for example, a rat that finds some new food never previously encountered and eats some of it. If the rat then falls ill, even an hour or more later, it will refuse to touch that food again; it will associate the taste with its subsequent illness, and condition an aversion to it. This is quite impressive. Traditional associative theories have always assumed that closeness in time is necessary for two events to be associated, but here is an association being formed in a single trial across an interval of an hour. And other laboratory tasks have revealed evidence of rapid conditioning across intervals, if not of an hour, then at least of several minutes. But this actually raises a further, quite different problem.

The world, even the world of a laboratory rat, is full of chance conjunctions of events. If the associative process can link two events that occur together once, even if separated by several minutes, why does it not do so always, leaving the animal's head filled with a whole host

Conditioning as selective attribution outcome

	t_1	t_2	t_3	
1	Coffee	—	Illness	Aversion to coffee
2	Coffee	Sucrose	Illness	No aversion to coffee
				Aversion to sucrose
3	—	Coffee	Illness	Aversion to coffee
	—	Coffee	No illness	
4	Sucrose	Coffee	Illness	No aversion to coffee
	—	Coffee	No illness	Aversion to sucrose

Figure 1 Schematic representation of four experiments on food-aversion conditioning: t_1, t_2, t_3 are three successive times (half an hour apart) when a rat may be given either some novel flavoured solution to drink, or an injection of lithium chloride

of irrelevant associations? The answer is that conditioning occurs selectively with better predictors registering at the expense of worse ones. We can illustrate this by some artificial abstractions from the real world designed to mimic the rat eating food that disagrees with it (Figure 1). In case 1, a rat is given a particular flavour to drink, here coffee, at time 1, and at time 3 (an hour later) is given an injection of lithium chloride which makes it mildly ill with stomach ache for half an hour or so. This is enough to condition an aversion to coffee, and the next day the rat will be reluctant to drink coffee. But this is only because coffee was the last novel-tasting substance it consumed before feeling ill. If, in case 2, we give the rat sucrose solution to drink at time 2, half an hour after the coffee and half an hour before it becomes ill, the illness is attributed to sucrose and not to coffee. In case 3, on some days the rat drinks coffee at time 2 and receives an injection of lithium half an hour later; on other days, it drinks coffee, but no lithium injection follows. The rat will then condition an aversion to coffee, even though coffee does not always make it ill, since it is still the best available predictor of its stomach ache. If, in case 4, the experimenter provides a better predictor – sucrose at time 1 on days when coffee is followed by lithium – the rat attributes its illness to the sucrose, not the coffee.

In all these cases the rat also takes its past experience into account.

In case 2, where it would normally attribute its illness to the sucrose rather than to the coffee, because that was the last thing it consumed before feeling ill, we can reverse this outcome in either of two ways. The day before the conditioning trial, we can either allow the rat to drink sucrose without adverse consequences or give it coffee to drink which does make it ill. In either case the rat will now attribute its illness on the conditioning trial to the coffee and not to the sucrose. It has remembered what happened the day before, and the conditioning process acts on the assumption that the world is a stable place. If sucrose was safe yesterday, or if coffee was not, then it is presumably the coffee and not the sucrose that is the cause of today's illness. The associative learning process thus seems remarkably well designed to allow animals to form an accurate picture of the stable causal structure of their world.

The rat in these experiments discriminates between different potential causes of its illness, and discrimination is, of course, an essential ingredient of intelligent anticipation. If food can be found in one place but not in another, simple conditioning will allow the development of an appropriate differentiation between the two situations. But it will also allow the development of the more complex conditional discriminations the world often requires: food is available in one place at one time of day, but only in a different place at a different time of day. In the laboratory, this situation can readily be mimicked by the experimenter. Rather than arranging a simple discrimination in which, say, a blue light always predicts food and a green light never does – so that the animal should always choose blue not green – the experimenter can decide that blue is correct only if the trial started with the illumination of a red light, while green is correct if it started with the illumination of a yellow light (Figure 2). If he is perverse enough, the experimenter can also set up a second-order conditional discrimination for his subjects. Red signals that blue is correct only if presented on a vertically striped background, not if it is on a horizontally striped background (and similarly for yellow). Pigeons, as well as primates, can solve all these problems. Provided that they can detect the stimuli being used, there is no evid-

Conditional discrimination

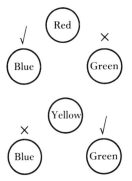

Second-order conditional discrimination

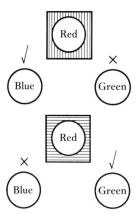

Figure 2 Examples of conditional and second-order conditional discriminations

ence that any of these discriminations is beyond the power of any vertebrate.

ABSTRACT REPRESENTATION

Associative learning thus provides animals, and people, with a powerful means of predicting the future, understanding what causes what, and discriminating between events of consequence. But does it

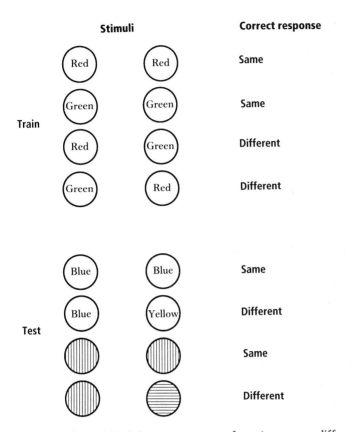

Figure 3 Training arrangements for a 'sameness–difference' discrimination. Having learned the training discrimination, subjects are tested for transfer to new pairs of stimuli

provide a sufficient account of animal intelligence? Surely not. The discriminations I have described, of which a pigeon is capable, although perhaps seeming complicated are susceptible to simple associative analysis. Other discriminations really do require a more abstract description of what it is that predicts the correct choice, and as proof of this, pigeons seem essentially unable to solve them (Figure 3). For example, supposing I train an animal on a 'Same/Different' discrimination: to make one response (labelled Same) to two circles of the same colour when both are red or both green, and another response (labelled Different) to two circles of a different colour. What the actual responses are is not particularly important. David Premack's chimpanzee Sarah will put one of two plastic tokens beside each pair she sees. A pigeon can be asked to peck at one disk for Same and another for Different. Irene Pepperberg's parrot Alex will actually say 'Same' and 'Different' – though we should probably not read too much into that!

All three animals will learn the problem, but they have not learned the same thing. Transfer tests suggest that Sarah and other chimpanzees have indeed learned to respond on the basis of sameness or difference, responding equally correctly to any new pair of stimuli. But the pigeons have not: they must learn each new problem from scratch as though it really were a new problem. The implication is that they learned the original discrimination simply by learning to make one response to two red or two green circles, and another response to a red and a green. Other primates will behave like chimpanzees, but this does not prove that we have found evidence of the phylogenetic scale I derided earlier; Alex the parrot also shows transfer – he can learn the relational rule, as can other birds such as crows and rooks.

Sarah can spontaneously sort objects on the basis of their similarity, which is perhaps more than a parrot would; but then Sarah has spent much of her life at this sort of task. Given four objects, two As and two Bs, and asked to put them into two piles, Sarah, like a three-year-old child but not like a two-year-old, puts the two As into one pile and the two Bs into another. Sarah can also sort on the basis of single attributes – and here it is clear that she is not alone, for so can Alex. Shown two objects differing in a number of respects and alike in only one, they can tell you which one that is. For example if Alex is shown a red circle and a green triangle, both made of wood, and asked 'What's Same?' he will give the correct answer: both wood. Sarah has been taught to sort on the basis of even more abstract attributes, such as fractions. Shown a half-filled jug of water and asked what goes with it – a half-disc or three-quarter disc – she selects the half disc.

There is reason to believe then that the perception of these sorts of relationships between objects in the world is an ability that both lies outside the scope of traditional associative learning theory and is beyond the capacity of a pigeon – and, I've no doubt, of a chicken or goldfish too. The reader may want to ask whether it is really a matter of superior intelligence, as we normally understand this term. My first response would be that it seems to be much more useful to point to a specific difference of this sort between, say, a pigeon and a crow or a parrot, than to say that crows and parrots are more intelligent than pigeons. That assertion seems, by comparison, wholly unilluminating.

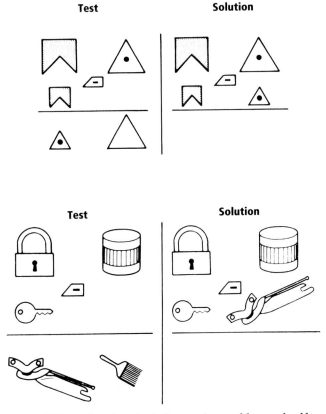

Figure 4 Examples of analogical reasoning problems solved by Sarah the chimpanzee. In each test, she is required to choose from the two alternatives below the line the one which ensures that the relationship between the right-hand pair of objects is the same as that holding between the left-hand pair. The solution is shown on the right. The objects in the lower example are a padlock and key, and a tin can and can opener

But my second response would be more emollient. Perceiving the similarity between two dissimilar objects or situations is precisely one of the means by which one can extend one's knowledge of the world to new areas. It is what we call seeing analogies, or what an IQ tester would call analogical reasoning. And Sarah the chimpanzee has no difficulty solving analogies of the sort: A is to A′ as B is to B′ (Figure 4).

CAUSAL INFERENCE

What else lies outside the scope of simple associative learning? No doubt many other things. Associative learning may allow one to predict effects from observations of their causes. But it will not solve all forms of causal inference. Consider the following experiment (Figure 5). With a chimpanzee watching through a screen, a trainer appears with an apple and a banana, places the apple in Box A and the

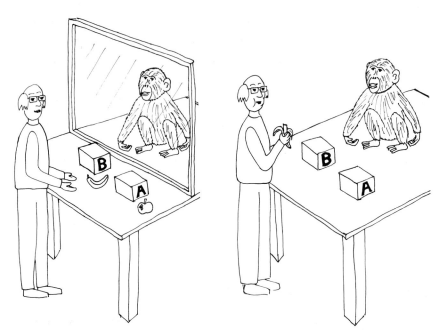

Figure 5 The initial phase (on left) and test phase (on right) of David Premack's experiment testing chimpanzees' ability to draw inferences

banana in Box B. The chimpanzee is then distracted for a moment and when she looks again there is the trainer guzzling the banana. She can now choose between the two boxes, but which will she choose? The answer is that she chooses Box A (the one with the apple). Why? Well, she has presumably inferred that the trainer has removed and eaten the banana he placed in Box B, and that Box B is therefore now empty. This is an inference – from the sight of the trainer eating the banana to the conclusion that the banana box is empty. There is, of course, no necessary connection. The chimp has not seen the banana being removed and indeed, it was not; the trainer had another banana in his pocket and it was this that he was eating. Nor was this a case of laboriously learning a conditional dis-crimination – if the trainer is eating a banana, then choose Box A because choice of Box B is not rewarded. It was an immediate infer-ence: Sadie was correct on the first trial and hardly made a mistake. This is quite impressive: we do not know if other animals could solve this – certainly, not all David Premack's chimpanzees did. Nor do chil-dren below the age of four. But four-year-olds do. At a later age, how-ever, they may become indifferent in their choice – reasoning appar-ently that the teacher or experimenter must be eating a different banana from the one she put in the box. And at least one ten-year-old, my son, said he would always choose the banana box because, being of a suspicious and pessimistic turn of mind, he assumed that if the experimenter was horrid enough to eat the banana, he would also be horrid enough to take the apple out of the apple box and put it in the banana box. Which just shows that some subjects can be too devious for a simple psychologist's experiment – or, more relevant to my argu-ment, that there is a significant gap still separating chimpanzee from human reasoning.

SOCIAL INTELLIGENCE

The examples of animal intelligence I have cited so far, it seems reas-onable to argue, are of obvious adaptive significance. But a question that has exercised many ethologists and comparative psychologists is why our close relatives, the primates, should be so very much more intelligent than any other animal. After all the great apes at least seem to live a pretty easy life. What are the selection pressures that have forced the development of intelligence in primates? One possible answer to this conundrum, of course, is to deny the truth of the premise – maybe primates are not particularly intelligent. A more popular answer has been that they live in large and complex social groups. In spite of the obvious objection that they certainly do not all live in such groups, and that there is no good evidence that the relatively solitary orang-outang is less intelligent than the chimpanzee, the argument is interesting, has a certain plausibility, and is surely worth taking seriously. After all, if one function of intelligence is to predict what will happen next, and if an important part of what will happen to you next is dependent on what someone else does, then you need also to be able to predict their behaviour. But if the complexity of their behaviour, and thus the difficulty of predicting it, increases as they become more intelligent, so your intelligence must increase to keep pace with theirs. And then theirs must increase to predict the increasing complexity of yours. And so on: society could have, so to say, a synergistic effect on the evolution of intelligence.

IMITATION

Certainly most primates are social, and since young primates typically remain dependent for a relatively long time, play with other young and so on, it seems plausible to suppose that they must learn a lot from one another. Do they do so by observation and imitation? We certainly assume that primates are pre-eminent among animals at imitation. English is by no means the only language where to 'ape' means to imitate. There is surely no doubt that apes will copy the actions of another. But so will other animals – many of the best examples coming from work with birds. And the importance or prevalence of imitation in monkey and ape societies has almost certainly been exaggerated. When one Japanese monkey learned to take the sweet potatoes thoughtfully left out on the beach by the research

workers who wanted to observe the monkeys' behaviour, and washed
the sand off in the sea, the habit spread to other members of the
troop. This was confidently interpreted as imitation, and an example
of 'proto-cultural' behaviour. One might remark, however, that if one
young monkey could learn to do it for herself so might others, and
that if it was a matter of imitation it was remarkably inefficient. Three
years after the original inventor had learnt the trick, less than 50 per
cent of her fellows had copied her, the remainder still settling for
eating gritty potatoes. Another example is that of chimpanzees 'term-
iting': this involves selecting a suitable twig, stripping it of leaves, licking it, inserting it into the termites' mound, wiggling it up and down,

Figure 6 A female chimpanzee fishing for termites, intently
watched by her infant

drawing it out and licking the termites off the tip. Surely, one ima-
gines, the infant so intently watching his mother perform this skilled
task (Figure 6) is learning what to do and how to do it. Well yes, he is
surely learning something – that food comes out of this unpromising
lump of earth – and that twigs have a lot to do with it. But, in fact, it
takes the young chimpanzee many years of practice by himself after
he has left his mother before he is any good at the task, and people
who have tried doing it themselves report that this is exactly what they
need. You can't really learn very much about how to perform the trick
just by watching someone else doing it.

COMMUNICATION

Social animals not only want to predict one another's behaviour, they
also seek to influence one another, to communicate information to
one another. But here we need to be particularly vigilant to guard
against overinterpretation. For example, vervet monkeys, who live in
large groups in the African savannah, give alarm calls when they
detect a predator – a different one when they detect a leopard from
that which signals they have spotted an eagle. So their calls are not
just undifferentiated expressions of fear, they convey more specific
information – which has specific effects on the behaviour of their fellows, who climb a tree when they hear a leopard call, and look up
at the sky when they hear the eagle call. But many species of birds
(starlings, chickens among others) also have differentiated alarm
calls for different predators which have different effects on their fellows. So there is nothing particularly special about the behaviour of
the vervet, and certainly nothing beyond the scope of simple associative learning theory. What might persuade one otherwise? Well, a
monkey or a chicken by himself is less likely to emit an alarm call than
one surrounded by fellow conspecifics. This has suggested to some
observers that alarm calls not only serve a social function, they are
intended to do so. The caller intends to warn his mate or offspring
or, even more altruistically, any other members of his species. And
since males are more likely to give alarm calls than females, a suffi-

ciently fanciful observer can detect a touch of male gallantry in his actions. But there is a simpler explanation which at least bears thinking about. A solitary chicken sees a hawk flying overhead. If he does not wish to be eaten, he has basically two courses of action open to him. He can freeze – crouch down immobile where he is – since it is movement that the hawk's eye is designed to detect first; or he can take a chance and make a run for cover – provided that there is cover close by. The first is usually the better strategy but it may not be if there are other birds nearby – especially if they are still running around because they have not seen the hawk. So our male observer is in a quandary. Is it better to sit still or run? He is in an even greater quandary than a female, since being more brightly coloured he is more likely than she to be spotted by a hawk alerted by initially seeing movement. But a common outcome of this sort of conflict between two response tendencies is that the animal does something quite different from either – in this case he starts squawking. This is what classical ethologists call a displacement activity, and a plausible account of the origins and elaboration, that is evolution, of a variety of calls and displays used in animal communication is that they first arose as displacement activities in situations of conflict. Alarm calls may be no exception.

So we have two quite different interpretations of alarm calls. One is that the caller intends to warn his fellows; the other that such calls are 'no more than' reactions elicited by a state of conflict. How might one decide between them? Well, if the caller really intends to communicate information to his recipient, he should take account of his recipient's prior knowledge. There is little point in telling someone something they already know. If everyone around you has already seen the hawk, there is not much point in shouting 'Hawk!' Not only chickens fail this test; so do monkeys. Here is one example. A female monkey with her infant beside her is allowed to look into an arena where she sees either some food or a predator being hidden. The infant alone is then allowed into the arena. If the infant had not been able to see what was being hidden, the mother makes the appropriate call. And the infant then starts behaving appropriately; cautious if he hears her alarm call, exploratory if he hears her food call. So far so good. But the mother behaved in exactly the same way when the infant was right beside her during the observation period, with at least as good a view as she had, and is already showing appropriately cautious or exploratory behaviour. The mother's calls are wholly unaffected by her infant's knowledge.

Now one might argue that some anxious human mothers would behave in a not dissimilar way, but this is not the only test that suggests that monkeys are quite insensitive to the difference between someone who knows something and someone who could not have such knowledge. For example: a monkey sees an experimenter hiding food in one of two boxes – but cannot see which. When allowed to choose, he will learn to go to the one the experimenter points to. Again, so far so good. But he cannot learn to tell the difference between an experimenter who hid the food himself and points at the correct box, and one who was not present when the food was being hidden, does not know where it is hidden and does not point correctly. The monkey is just as likely to go to the box which the necessarily ignorant experimenter points to – even if we rig the experiment in the monkey's favour by ensuring that the ignorant experimenter actually always points to the empty box. The monkey just does not learn this conditional discrimination – although a chimpanzee can.

We should not necessarily be surprised at the monkey's failure – although we should probably be impressed by the chimpanzee's success. According to some developmental psychologists, children do not really understand what it is for someone else to know or not to know something until they are about four years old. Until this age they fail the 'false belief' test first devised by Heinz Wimmer and Josef Perner. In a typical version of this test (Figure 7), the child is told a simple story enacted by two puppets. The first puppet, Sally, places her marble in a basket, and then leaves the room. The second puppet, Anne, then removes the marble from Sally's basket and hides it in her own box. When Sally comes back into the room the child is asked where she will look for her marble. Four-year-old children correctly answer that Sally will look in her basket – where she put it. Even though the child

Figure 7 Schematic representation of the 'false belief' test. The child (C) sits opposite the experimenter (E), and in (1) watches the experimenter get Sally to put her marble in the basket

knows that the marble is in Anne's box, Sally cannot be expected to know this, since she was out of the room when Anne moved the marble. But three-year-olds claim that Sally will look, where they know the marble to be, in Anne's box. They fail to attribute to Sally a belief which they know to be false.

The implication drawn by some developmental psychologists from this test is that, below the age of four, children do not attribute knowledge or belief to others. That conclusion has been disputed by other psychologists who have argued that other diagnostic tests reveal that even two-year-old children understand that others have beliefs. Their favourite test is lying or deception.

This brings us back to the world of monkeys and apes, for in recent years a large, richly anecdotal literature of tactical deception of one monkey or ape by another has grown up. Deception does indeed seem a good test case, for if I lie to you, I am trying to get you to believe something which I believe to be untrue. So lying seems to imply that I believe that you have beliefs. Does the deceptive monkey or small child believe that he is manipulating his dupe's beliefs? The trouble is that lying also has effects on the behaviour of others. The small boy who says 'It wasn't me who did it, it was Susie' is trying to avert his mother's wrath, or perhaps trying to get his sister into trouble. And the fact that a two-year-old will say this even when Susie was not in the room at the time, and his mother was and clearly saw what happened, suggests that there is something seriously amiss with his understanding of what it is for someone else to know or believe something. If so, perhaps we should settle for a simpler explanation of his lie: he was not trying to influence his mother's beliefs, only her behaviour. But this is a crucial difference, since someone who tells a successful lie will usually succeed in manipulating another person's behaviour to their own advantage. And simple conditioning theory is thus quite sufficient to explain why we, or other animals, learn to deceive one another. To show that such deception is intended to influence the dupe's beliefs, we should want to know that the animal passes the test that the two-year-old boy failed: do not tell a lie if your intended dupe clearly already knows the truth.

I do not dispute that an animal who manipulates the behaviour of another is engaged in a complex social interaction; the deceiver must carefully monitor the dupe's behaviour and be able to predict what he will do next. This calls for a high level of social intelligence – but we should not exaggerate the level of intelligence called for by attributing beliefs about beliefs, without very much better evidence than we have so far. There is already enough exaggeration without bringing that in. For many cases of apparent deception do not even require us to suppose that the deceiver intended to influence the behaviour of the dupe. A story repeated of chimpanzees both in captivity and in the wild is that after a quarrel with another chimpanzee or with their trainer, they will make friendly overtures, enticing the trainer over with gestures of friendship and reconciliation. The trainer approaches – only to get thumped when in range. What more natural than to say that the chimpanzee deceived the trainer, pretending to be friendly but all along planning her revenge? Well, perhaps. But it is equally plausible to see the chimpanzee as in a state of conflict between positive and negative feelings, oscillating between friendship and aggression, affection and fear. When the trainer is far away the positive feelings win out, when close up the negative feelings win. That sounds far too pat. But it is exactly what one would expect from other studies of conflict between approach and avoidance. Confronted with an object associated with both positive and negative consequences, the laboratory rat approaches with ever increasing hesitation, only to dart away again if he gets too close. Attraction generalises more widely than aversion; at a distance therefore the attraction wins out, close up the aversion may. When her trainer was far away, the chimpanzee felt friendly; when he came too close, she felt aggressive, so she hit him. But there was no intended deception at all.

CONCLUSION

My conclusion is that animals are both cleverer and more stupid than we think. They are cleverer because they can learn to solve problems that seem much more complex than we might have supposed; they are more stupid because we are far too inclined to attribute to them more complex mental states than their behaviour actually warrants. We do not see the contradiction because we make one error about some animals, the other about others. We are surprised to learn that a pigeon or chicken can solve complex discrimination problems, which we would certainly not be surprised to see solved by a monkey or ape. It is their behaviour we are all too happy to interpret in the richest possible way. And this is because we are unable to rid ourselves of a third misconception, that there has been a linear increase in intelligence as we progress up the phylogenetic scale. Of course, there are some sorts of problem a chimpanzee will solve which a pigeon will not (and vice-versa). The job of the comparative psychologist is to elucidate the particular processes underlying such differences – not to attribute them to a global difference in intelligence. That is both too easy and too uninformative. Indeed, I could make a strong case for the suggestion that we should eschew all use of the term 'intelligence'. But that might get me into trouble with the other contributors to this book.

FURTHER READING

Byrne, R. W. and Whiten, A. (eds.), *Machiavellian Intelligence: Social Expertise and the Evolution of Intellect in Monkeys, Apes and Humans*, Oxford: Oxford University Press 1988.

Gallistel, C. G., *The Organization of Learning*, Cambridge, MA: MIT Press 1990.

Mackintosh, N. J., *Conditioning and Associative Learning*, Oxford: Oxford University Press 1983.

Parker, S. T. and Gibson, K. R. (eds.), *'Language' and Intelligence in Monkeys and Apes*, Cambridge: Cambridge University Press 1990.

Premack, D. and Premack, A. J., *The Mind of an Ape*, New York: W. W. Norton 1983.

Ristau, C. A. (ed.), *Cognitive Ethology: The Minds of Other Animals*, Hillsdale, NJ: Lawrence Erlbaum Associates 1991.

Whiten, A. (ed.), *Natural Theories of Mind*, Oxford: Blackwell 1991.

3

Infant intelligence

GEORGE BUTTERWORTH

Scientific psychology draws upon a variety of methods in attempting to explain intelligence. This essay will address the problem of explaining human intelligence from the theoretical and methodological viewpoint of developmental psychology. The developmental – or genetic – method, attempts to explain intelligence by observing its origins and how it grows. Questions about human intelligence and the mechanisms, processes and circumstances under which human intelligence is originally expressed naturally lead to a particular concern with infancy. A number of issues will be raised about traditional conceptions of intelligence, in the light of contemporary evidence about how babies perceive and understand their physical and social world.

My argument falls into three parts. First, I look at the most widely accepted theory of the origins of human intelligence, that of Jean Piaget. This provides the necessary background to illustrate some universal features of human intelligence, which Piaget suggested have their origins in patterns of action. Having set the scene, I examine, second, contemporary evidence for sophisticated perceptual abilities in very young infants, which casts some doubt on Piaget's theory. Finally, I review some recent work on predicting individual differences in intelligence from infancy to early childhood.

Before embarking on this review, however, it is necessary to say something about the definition of intelligence. Rather than simply defining intelligence as a single, underlying intellectual ability, we will also take into account aspects of the physical and social context. Three criteria seem essential to the definition of intelligence:

- Intelligence cannot be well understood without reference to the internal representation of knowledge. Using a knowledge base to decide what information is relevant and what irrelevant is a feature of intelligent behaviour. One way to understand intelligence therefore may be by reference to the psychological processes which give rise to knowledge.

- External, contextual factors as well as internal, cognitive factors enter into the structure of intelligence, since what may be intelligent in one context may be superseded or become unintelligent in another. So, a second criterion for the definition of intelligence may be by reference to behaviour in a social and physical context.

- Finally, it may be necessary to distinguish between how intelligence is expressed in behaviour – which may indeed be very limited – and what babies may be capable of perceiving, which may be unsuspectedly sophisticated. That is, infants may not always reveal underlying intellectual competence where their means of expression are limited by motor immaturity.

With these criteria in mind the following working definitions, one rather old and the other relatively recent, can be adopted: 'Intelligence (Intellect) is the faculty or capacity of knowing' (J. M. Baldwin, 1905), and 'Intelligence is expressed in adaptive behaviour, in particular contexts' (R. J. Sternberg, 1982).

INTELLIGENCE IN BABIES

Among the earliest systematic observations of babies were those of Charles Darwin, who had actually observed the development of his own infant son, Doddy, in 1840. Darwin observed various reflex actions during Doddy's first 7 days, such as sneezing, hiccoughing, yawning, stretching and crying. Darwin observed the first evidence of 'practical reasoning' in his infant son at the age of 114 days, when the baby slipped his hand down Darwin's finger in order to put the finger in his mouth. Darwin also remarks on the early age, at or before 4 months, when Doddy showed curiosity and imitation.

Most subsequent studies of very young infants emphasised the reflexive organisation of early behaviour. For example, the behaviourist James Watson, investigated the grasping reflex in newborns in 1921. He was able to show that the infant could suspend her own weight through the grasping reflex, so tightly did she grip the support. The implication of this demonstration, however, was that development originates in a rather limited 'unintelligent' repertoire of inherited automatic reflexes.

Piaget's observations of cognitive development in infancy stand out as among the most thorough attempts to explain the origins and development of human intelligence. Piaget based his theory on the work of the nineteenth-century psychologist James Mark Baldwin, whose definition of intelligence was given earlier. According to Piaget, intelligence has its own developmental history. Piaget agreed with Darwin and Watson that the process of intellectual development begins from the most elementary reflexive adaptations. From this limited set of innate reflexes, such as sucking and grasping, there develop, during infancy, basic programmes of intelligent action, known as circular reactions, which eventually give rise to verbal intelligence and thinking.

Stated in the most general way, Piaget's theoretical assumption was that the roots of intelligence lie in programmes of action which form the first link between the baby and the world. He was concerned to explain the universal aspects of development of human intelligence, rather than individual differences between people. This essay will be structured in a similar way, first concentrating on some universal aspects of human intelligence; only then will the origins of differences between individuals in intelligence be considered.

PIAGET

According to Piaget intellectual development in infancy proceeds in a series of stages, as the baby progressively adapts her actions through encounters with objects. Piaget set out to explain how intellectual development arises from biological roots, of which the most fundamental hereditary mechanisms are the innate reflexes. Figure 1 summarises Piaget's theory of sensori-motor development. He argued that each use of a reflex changes the underlying control system in such a way that the reflex incorporates earlier experiences. The infant soon applies the reflex to as many objects as she possibly can. Repetition of an acquired cycle of activity, for example thumb sucking, is known as a primary circular reaction. The circular reactions are so called because they are self-sustaining, repetitive activities whereby the action itself, through its sensory consequences, is the stimulus for its own repetition. Coordinations among circular reactions, at successive levels of organisation of development, give rise to various forms of knowledge and hence structure the intellect. The coordination and hierarchical integration of circular reactions are the means whereby the intellect eventually transcends its biological roots.

Figure 1 shows how intelligence proceeds from innate roots in reflexes, through habitual forms of action, to the intentional use of means to achieve certain ends. The first intentional actions may be observed at around 9 months, along with the capacity to search for hidden objects. The first insightful solutions to problems may be observed with the further development of memory between 12 and 18 months. The end of infancy is marked by the acquisition of language and mental imagery at about 18 months.

In summary, according to Piaget, intelligence is expressed originally as programmes of action. Action organises and gives structure to perception and, eventually, it lends structure to thought and language. Piaget attempts, in a rather sophisticated way, to account both for continuity – with an innate biological endowment – and for discontinuity – as intelligence develops and acquires new properties.

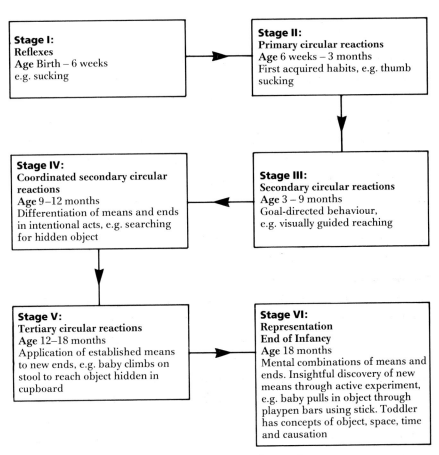

Figure 1 Piaget's hierarchical theory of sensori-motor development

PROBLEMS WITH PIAGET'S THEORY

It would seem, therefore, that universal aspects of human intelligence can be explained by applying the genetic method, as Piaget did, to trace intellectual development to its roots. But like any theory, Piaget's account stands or falls on his starting assumptions and it is here that much recent research has revealed unsuspectedly sophisticated abilities in young babies.

The fundamental problem concerns his assumptions about the initial adaptation of the infant to the environment. According to Piaget, perception in the very young infant is inadequate to inform the baby of the objective properties of reality. Piaget distinguished perception and intelligence, following Baldwin's definition of 1905: 'Intelligence (Intellect) is the contrast between sensations as material on the one hand and the mind's elaboration of this material on the other'.

The traditional assumption is that perception lacks structure; it consists of mere sensations, which must be interpreted and given coherence and meaning by the mind. On this view, perception in the newborn is inadequate to inform the infant of a three-dimensional, spatially extended world. Indeed, according to Piaget, visual perception in the early months of life is of two-dimensional 'tableaux'. On this theory it is only by coordinating touch with vision that visual objects acquire solidity and only by moving objects about in the field of view, by crawling and eventually by walking, does space become separated into various planes of depth. It falls to the motor system and to the sense of touch to tutor the visual and auditory senses. Since all knowledge originates in action, infant development consists in a gradual awareness of an extended and distant space. Development is from proximal to distal knowledge as the baby first masters her own body and then discovers the world. Piaget's theory therefore leads to the view that the mechanisms of distance perception – visual and auditory perception – are only of secondary importance in attempting to explain intelligence.

GIBSON'S 'GROUND THEORY'

The theory that the infant derives all knowledge from action began to come under suspicion with the work of DeCarie who showed that thalidomide babies, some of whom were born without arms or legs, nevertheless passed through Piaget's sequence of sensori-motor stages at approximately the correct ages. Most of these babies developed normal intelligence, thought and language, despite their motor handicaps which deprived them of experiences, such as coordinating vision with touch, which are supposedly critical for intellectual development.

However the major impetus for a re-evaluation of the importance of early perception came from the theoretical writings of James Gibson on the mechanisms of space perception. Gibson asserted that perception in the newborn does not begin with a flat patchwork of visual sensations to which depth must be added by learning. Particularly important for understanding this point is Gibson's 'ground theory'. According to Gibson, visual space is perceived through dynamic changes in the pattern of light reflected from the textured surfaces formed by objects and by the ground. Systematic changes in the flow of textured information are specific to the movement of objects in the world or to movements of the observer in a three-dimensional visual space. This is illustrated in Figure 2 which simulates the texture flow pattern that occurs when an aeroplane is brought into land. Similar flow patterns also occur when we move through space or when we move our head or eyes. Systematic disruptions of textured flow patterns occur when objects move about in space and when they pass in front of or behind each other. The idea, as we shall see, is that information derived from the dynamics of perception may be sufficient to inform even the youngest baby of her relation to the visual world and the objects that move about within it.

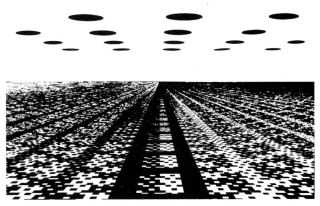

Figure 2 Example of a flow pattern when an observer is in motion relative to a textured visual environment

On this view, then, perception is adequate to the task of detecting the relation between the infant and the environment. The essential points for theories of the origins of intelligence and intellectual development are, first, that perception depends upon the detection of the fixed and varying properties of the environment, rather than on their construction through action; and, second, that babies might be expected to attend to those properties of the environment that hold adaptive significance for them. Taking such a view seriously leads us to ask whether intelligence may not have its roots in the mechanisms of perception and attention with which we are innately endowed.

CONTEMPORARY EVIDENCE

Evidence for great sophistication in infant perception has accumulated at an enormous rate since studies were first carried out in the early 1970s. If infant perception were merely of two-dimensional visual tableaux, and consisted of unconnected sights, sounds, touch and smells, as was traditionally supposed, then it would be impossible for the baby to perceive the world of objects as do adults. There have been many studies which show that this is simply the wrong way to characterise the perceptual world of the newborn. For instance, babies less than 3 months old perceive a visual object, such as a rectangle, as having a constant shape and size, despite differences in the distance and angles through which the object is presented. That is, the visual perceptual system operates in infants who cannot yet crawl or walk, according to principles of object constancy. People and things are perceived as having a particular shape, size and distance in space.

Other important evidence concerns the relations between vision, hearing and touch, which were traditionally thought to become coordinated only in the sixth month of life. A very simple technique, known as the visual preference method, is used. Babies are given a pair of displays, on left and right of the field of view, and their preference between the displays is recorded. The fact that babies have preferences itself shows that they actively seek after visual information;

they are not simply passive recipients of visual stimulation. By clever manipulation of the displays it is possible to learn a lot about the mechanisms of perception in very young babies.

In the case of vision and hearing, a great deal of evidence has accumulated to show that even the youngest baby may be in a position to relate sights to sounds. One of my own studies for example, showed that babies just a few hours old will turn their eyes toward a sound played to one ear. This shows that audition and vision are at least to some degree, coordinated from birth.

More recent studies have shown that newborn babies are sensitive to the patterning of sights and sounds. For example, Elizabeth Spelke has shown that newborn babies prefer to look at whichever of two identical videos, viewed side by side, is synchronised with the sound track played over a single loudspeaker between the two television sets. In this study, both the films showed an object moving up and down in a vertical movement but only one film was in phase with a regular noise as the object made contact with the floor. Infants preferred the display where sight and sound were synchronised.

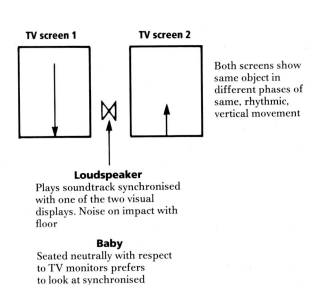

TV screen 1 **TV screen 2**

Both screens show same object in different phases of same, rhythmic, vertical movement

Loudspeaker
Plays soundtrack synchronised with one of the two visual displays. Noise on impact with floor

Baby
Seated neutrally with respect to TV monitors prefers to look at synchronised display

Figure 3 Method for showing infant preference for synchronised sights and sounds

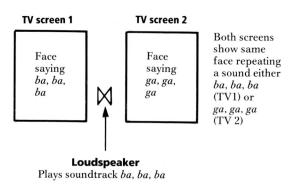

Figure 4 Method for showing infant preference for synchronised faces and voices

The innate coordination between seeing and hearing has important implications for the acquisition of speech, as Kuhl and Meltzoff have shown. Babies of 3 months prefer to look at one of two films showing an actor mouthing the sound *ba, ba, ba* coincident with the soundtrack, rather than toward the same face mouthing the sound *ga, ga, ga* where the mouth movements are inconsistent with the soundtrack. The research implies that babies may be able to make use of lipreading to assist them in the perception and production of speech.

Similarly unexpected findings have been made with respect to the relation between vision and touch. Meltzoff and Borton, for example, give babies of 28 days an unusual, knobbly dummy to suck, without the infant being allowed to see the dummy before it was placed in the mouth. They then showed the babies large models of the knobbly dummy on one side of the visual field and a smooth dummy on the other. Babies preferred to look at the dummy with the same texture as they had previously sucked. It seems that oral touch in the very young baby conveys information about texture and/or shape which transfers to the visual system. In other words, what something feels like in baby's mouth provides some information about what it looks like. There are many other examples which lead to the conclusion that perceptual systems in babies have evolved to pick up information which attests to reality. So, the perceptual world is not the incoherent collection of isolated sensations, characterised in William James' famous description of the world of the newborn, as a 'buzzing, blooming confusion'.

It can be concluded that the relations among the senses are not as undeveloped as the Piagetian account had assumed. It could be the case that both the universals of human intelligence, and individual differences among people, originate in mechanisms of perception.

OBJECT PERMANENCE

The remainder of this argument will be pursued in the context of a limited number of examples, all of which relate in one way or another to infants' perception of physical and social objects. One of the central problems to be addressed is how we know that an object continues to exist even when it leaves our immediate visual experience, as for example when one object passes behind another. On the traditional account, out of sight is out of mind, since the object itself no longer provides sensations once it passes behind a screen. According to Piaget, knowledge of object permanence is only slowly acquired. Before 8 or 9 months, babies fail to search for a hidden object because, as far as they can tell, the hidden object has ceased to exist.

The Belgian psychologist Albert Michotte offered an alternative account for how we perceive permanence based on the dynamics of the visual image when an object disappears behind a screen, or when it is plunged into darkness. Other kinds of transition specify the annihilation of the object, as for example, when a puddle of water evaporates. The dynamics of the visual image, as the object moves from in sight to out of sight, contains the information for the continued existence or annihilation of the object. His explanation was 'When I speak of a prefiguring of ideas of causality, permanence and so on (in perception) I have never tried to pretend that these ideas are preformed in the sense of classical, innate ideas. What I have in mind is the existence of intrinsic signs of certain mental structures . . . One can perceive something slide behind another object without having to have the concept of a permanent object.' In other words, Michotte wished to establish a clear distinction between perceiving that an object continues to exist after it vanishes and knowledge that objects are permanent.

Various ingenious techniques have been devised to show that young babies do perceive occlusion as 'one thing going behind another', even though they will not search with their hands for hidden objects. Thomas Bower was the first to test whether babies perceive occlusion. Infants aged 7 weeks were trained to suck on a nipple in the presence of a large red ball. Once they had learned to do this, the ball was made to disappear by slowly drawing a screen in front of it. Babies continued to suck after the object passed behind the screen, suggesting that the learned response to the ball was maintained because they perceived it to be 'present' but invisible. However, when the ball was made to disappear instantaneously, by an arrangement of mirrors, so that it looked as though it had suddenly been annihilated, the babies ceased to suck. Bower interpreted this to show that the baby perceives that the red ball no longer exists and so the learned sucking response to the object becomes inappropriate.

Another, widely used technique for studying infant perception is known as habituation. This involves showing a baby a particular display until the infant begins to become bored with it. Once habituation has occurred, usually when the infant's interest in the display drops to one half its initial level, a change in the display is introduced. Recovery of attention occurs if the baby has registered the difference between the new display and the old one. (To anticipate the conclusion, it is worth mentioning that the habituation test proves to be an important predictor from infancy of individual differences in intelligence.)

Kellman and Spelke used the habituation technique to establish whether 4-month-old babies perceive partially hidden figures as whole objects or as disconnected parts. They showed that a display

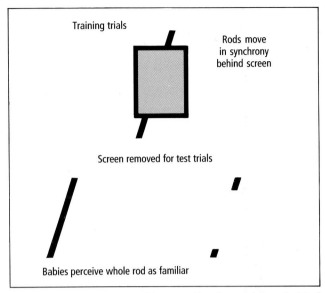

Figure 5 Infants perceive objects moving in synchrony behind
a screen as a whole object

comprising two separate objects moving in synchrony behind a
screen was perceived as a whole object whose centre was covered by
the screen. Babies revealed that they were expecting one whole object
behind the screen, because when the screen was removed to reveal
two separate objects they showed renewed interest in the display.
When the screen was removed to reveal a whole object, they remained
habituated to the display, with no recovery of attention.

Computer technology has been used to create the appearance of
depth on a television screen. A vertical column of dots is made to move
in such a way that dots in the remainder of the array are erased as
the vertical pattern moves over them. To the adult eye, the deleted
surface appears to be behind the moving surface. The wiping over of
one part of a visual display by another is sufficient, for adults at least,
to perceive a boundary in depth between the two surfaces. Babies
aged 5 months, when shown such a computer-generated display, preferred to touch the surface that was apparently nearer to them on the
monitor (Figure 6). The 5-month-old babies, just like adults, perceive
depth at the shearing edge of the moving visual display. Information

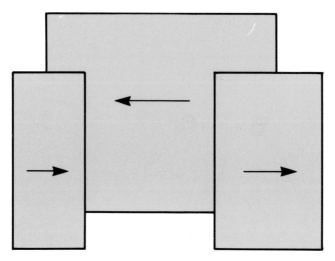

Figure 6 Texture deletion and the perception of surfaces in depth. Arrows show direction of motion

from relative movement therefore seems to be used reliably to inform the infant that two objects are separated in depth and this obviously carries the implication that the occluded surface is present but invisible.

Another technique which involves perception of occlusion shows that babies of 4 to 6 months can pick up complex information concerning human locomotion, from minimal perceptual input. The technique involves computer models of so called 'point light walkers'. Point light walkers are created by placing lights or luminous tape on the head, torso and limb joints of a person dressed in black, who is then filmed while traversing a path at right angles to the observer's line of sight. Adults viewing the filmed dots in motion report a compelling experience of seeing a human figure walking. Infants prefer to look at a display showing this biological motion than a display where the same number of dots simply move randomly. Detailed analysis has shown that the earliest dynamic information which the baby picks up involves occlusion when the lights on the hip joint and the far limbs of the body are hidden in a regular succession by the lights nearer the observer.

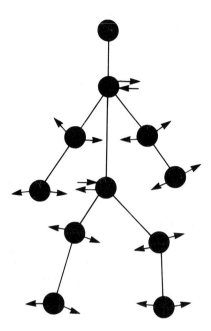

Figure 7 Moving points of light attached to head and major joints of a walking person

Other tests confirm that babies do indeed perceive the permanence of objects. For example, the baby is shown an interesting object, within reaching distance. The lights are turned off and the baby's reactions are filmed using infra-red photography. Infants of 5 months will reach in the dark, in search of the object. Out of sight is not out of mind; babies use visual information to inform them of a world of spatially connected, separately moveable, whole, permanent objects in the first 5 months of life. Yet, they don't search with their hands for objects hidden under a cloth before 8 or 9 months.

In fact, even after they search manually, they make curious errors which led Piaget to suggest that they do not understand that an object can only be in one place at one time. Piaget argues that babies perceive the existence of the hidden object still to be linked to their own action. An infant between 9 and 12 months will search for a hidden object at a place A, successfully. But if the same object is hidden at a new place, B, the baby will persevere and search again at A even though the object was seen to disappear at place B. For Piaget this was definitive evidence that action gives intelligent structure to perception in infancy and for the subordination of perception to action.

A great deal of research has been carried out on these so called perseverative errors in manual search and there is no doubt that such errors do occur. What is critical in the present context is to establish whether perseverative errors are inevitably made by babies, as Piaget's universal theory of intellectual development requires. In an extensive series of studies we systematically varied both the colour of the cover and distinctive features of the background. We found that the tendency to make perseverative errors varied systematically with the visual structure of the successive spatial locations at which the object was hidden. Indeed, under some visual conditions, when distinctively different covers were placed on a continuous background, the babies made no errors at all.

In other words, babies can identify an object across successive changes in position. For them to reveal this ability, however, the visual–spatial conditions of testing have to be just right. The spatial context must be stable and contain sufficient landmarks to enable the object's movements to be monitored.

So, why don't babies search for hidden objects? Is it because they do not perceive permanence, as Piaget argued, or could it be something to do with how action is organised? The evidence suggests that the difficulty does not lie in the disappearance of the object, as was traditionally supposed. In fact, the problem may lie in generating the sequence of actions needed to fulfill the goal of obtaining the object from behind the screen.

This is perhaps best illustrated by considering slightly older babies who could retrieve a toy car hidden beneath a cloth but who nevertheless failed to retrieve a complete visible object before about 13 months, when the task involved a small detour to one side of the line of sight. Nicholas Jarrett has shown that when a toy car is placed immediately behind the hole in a transparent screen, the baby has no difficulty in retrieving a car along her line of sight. However, as soon as the car was moved to one side of the hole, so that an indirect path must be traversed between baby and object, the infant was com-

Figure 8 'Detour' problems present great difficulties to babies

pletely unable to solve the problem. The child forcibly attempted to remove the screen, which was treated as a barrier to her perception and possession of the object (Figure 8).

Such 'detour' tasks show that the infant seems to lack the ability to organise an indirect pathway for the hand to move to the object. So we have evidence for intelligent perception in early infancy, which Piaget never suspected, yet he was correct to argue that problems which require indirect solutions, such as the detour task, give difficulty until well into the second year of life. Perception is not enough to guide some types of intelligent action, involving multiple sequences of means and ends.

SOCIAL ASPECTS OF COGNITIVE DEVELOPMENT

Having said something of the preadapted nature of object perception in babies the argument can now be extended to introduce the social context into this discussion of the perceptual origins of intelligence. How babies follow another person's line of sight may serve as a convenient vehicle to discuss the issue. The technique is simple: adult and infant are engaged in face to face interaction until, at a prearranged moment, the adult turns to inspect an object located somewhere in the room. Infants as young as 2 months of age will turn head and eyes in the appropriate direction, as if to discover the object at the focus of the adult's attention. This finding, originally made by Scaife and Bruner, was most unexpected since it had long been assumed that babies are totally egocentric and incapable of understanding that others also have points of view. The most straightforward interpretation is that the infant responds to the adult's signal as 'something there', a possible object potentially available to the infant. A change in the focus of visual attention of the adult acts as a signal to the infant that a potentially interesting object exists somewhere in the general direction of the adult's focus of attention. The importance of the phenomenon of shared visual attention is that it offers a means of exploring early processes of communication, before language, and perhaps gaining some insight into language acquisition itself. Joint attention and pointing are frequently accompanied by the mother verbally labelling the object of interest.

Jarrett and I have studied babies aged 6, 12 and 18 months in interaction with an adult, usually the mother. The adult was under instructions to gain the infant's close attention and then to look at one of several targets located at various positions in an otherwise empty room. The main findings were that babies in all three age groups could locate targets within their own visual field but rarely if ever localised targets behind them, outside their visual field. We interpret these results to mean that the baby perceives her own visual world to be held in common with others; the infant is not locked into an egocentric universe. On the contrary perception brings the infant into social contact and forms the substrate for communication.

It is of particular interest to link our research with the mechanisms controlling pointing in babies. Babies begin to comprehend another person pointing at about 12 months, and they will point things out for others at about 14 months. Pointing, with outstretched arm and index finger is a gesture that is species specific to human communication. Our most recent studies show that, by 14 months, the baby can accurately localise which of two identical objects the mother is pointing at even when they are only a few inches apart.

These studies therefore suggest that early referential communication may grow out of more basic processes of reference implicit in unlearned processes of perception which presuppose a world of objects. The social basis of joint attention may depend on signals which carry the information from the mother to the baby that something 'out there' is of mutual interest. It seems that the precursors of universal aspects of human social intelligence, such as are manifest in index finger pointing, may ultimately be rooted in processes of perception.

This argument can also be pursued from the level of universals of human intelligence to consider intelligence in its individual aspects. Research on caretaking interaction between mothers and infants shows that mothers who most frequently encouraged their infants to attend to objects or events at the age of 4 months had infants with larger vocabularies at 12 months. The researchers, Bornstein and Ruddy, called this style of interaction didactic encouragement. Alternative explanations, such as that babies who vocalised a lot elicited more attention from their mothers, and subsequently acquired larger vocabularies, is not supported by the evidence. The amount of vocalisation at 4 months did not correlate with vocabulary size at one year. It seems therefore that one of the many factors which contribute to the development of intelligence is the extent to which adults engage the attention of infants in mutual interest in the world of objects and this coupled with appropriate speech, may eventually show up in aspects of verbal intelligence.

PROFOUND MENTAL HANDICAP

Another way to approach the question of explaining intelligence might be to look in detail at the structure of intelligence in the mentally handicapped, again using Piaget's theory as a framework. We could think of this approach as a critical test, since those abilities most deficient among the least intelligent are perhaps those that would ordinarily be most implicated in normal intellectual development. Our hypothesis was that these profoundly mentally handicapped children might be considered as if they were fixated at an infantile stage of intellectual development. This hypothesis may be contrasted with the alternative view that profound mental retardation is the outcome of a developmental process that is fundamentally different from that observed in normal children. (Of course, a combination of the deficit and difference perspectives may prove to be the most appropriate).

Macpherson and I carried out an extensive study of the structure of intelligence in a group of forty-five profoundly mentally handicapped children aged from 3 to 18 years. All had symptoms of brain damage and were being cared for in two hospitals, some as day patients. To give some idea of their ability in IQ terms, profound and severe mental handicap is defined as an IQ below 50, a score which would be exceeded by 99 per cent of the population. The children were tested using the Uzgiris–Hunt scales of sensori-motor development, which are based on Piaget's description of development in infancy. The test has seven subscales including tests of object permanence, of understanding of means–ends relationships, of vocal and gestural imitation, of physical causality and of spatial development.

The test results for normal babies showed that development proceeds synchronously across all seven scales. At any particular age, performance on the tests was similar. However, among the profoundly mentally handicapped group development was not synchronous; there were deficiencies in object permanence and gestural and vocal imitation hardly developed beyond rudimentary levels. Looking at these extremely intellectually handicapped children as if they are babies who have become 'stuck' at an early stage of development, therefore suggests that they have particular difficulties with object permanence and with imitation.

These are potentially very important results because we have already suggested that permanence may be specified perceptually. It may also be the case that imitation depends ultimately on perceptual systems. Some very important studies by Meltzoff have shown that newborn infants can imitate mouth opening, tongue protrusion and simple manual gestures. The ability to imitate is based on perception of the equivalence between the act seen and the baby's own action.

Meltzoff argues that this requires the same ability as is involved in perceiving the correspondence of information between vision and touch. Meltzoff has also carried out research which has linked neonatal imitation to processes of lip reading and speech perception that may aid the normal infant gain articulatory control and normal speech.

Thus, if the structure of the intellect among the profoundly mentally handicapped provides a critical test of the abilities required for normal intellectual development, we may argue that the ability most impaired is that enabling the child to translate auditory, tactual or visual information into motor analogue form. The ability to perceive the equivalence of perceptual input and motor output, which is thought to form the basis of the recently discovered perceptual competence of the normal infant, seems to be precisely the ability which is most disturbed in the profoundly mentally handicapped child.

PSYCHOMETRIC MEASURES OF INTELLIGENCE

The discussion so far has mainly dealt with the concept of intelligence considered as a universal. Standard tests of infant development include many motor items, such as the age of achieving the major motor milestones, or gaining head control, sitting, crawling or walking. They also test motor skills such as stacking bricks and simple hearing functions. Try as they might, constructors of such tests have been unable to establish any strong correlation between these measurements taken in infancy and later IQ. The lack of prediction was taken to mean that infancy must be discontinuous with later intelligence tests.

In recent years a number of reports have appeared in which it is claimed that infants' differential visual fixation of novel stimuli in habituation tests validly predicts later verbal IQ. Alan Slater has carried out several longitudinal studies which relate visual attention measures between 3 and 18 months to IQ between 4 and 8 years. Infants are presented with complex visual stimuli, including faces or complex coloured scenes in habituation tasks. The duration of the infant's first look at the stimulus, the total amount of time spent looking and the average duration of fixation over trials correlated with verbal IQ, both at 18 months and in the early school years.

It is not yet known whether it is the ability to perceive, to encode, to abstract, to categorise or to retain information that is implicated in producing these significant correlations. Fagan and Singer claim that visual preference tests administered in infancy, when babies are aged between 3 and 7 months, predict about 20 per cent of the variability in childhood verbal IQ measured between 3 and 7 years. This level of predictive validity means that perceptual functioning in infancy cannot be the only factor involved in determining IQ. Nevertheless this is significantly better than the predictive ability of standard infant tests. It is also interesting, in the light of our previous discussion about joint attention and pre-verbal communication, that the verbal IQ tends to be predicted better than non-verbal IQ. The implication is that social processes are implicated in the development of intelligence, which are then revealed in verbal abilities. These detailed issues remain to be resolved. The important new facts are that measures of visual fixation provide the first evidence for continuity between infancy and later intelligence. This is consistent with our general thesis that the roots of intelligence may lie in processes of perception.

To conclude, if intelligence is to be explained, among the important factors that should be considered are how perception and attention may be intrinsic to its origins, structure and development. Perception should not be conceived of as subordinate to intellect, as has traditionally been the case. This insight has important implications for our understanding of the origins of intelligence in the normal infant.

FURTHER READING

Bower, T. G. R., *Development in Infancy*, 2nd edn, San Francisco: Freeman 1982.
Bremner, G., *Infancy*, Oxford: Blackwell 1988.
Butterworth, G. E. (ed.), *Infancy and Epistemology: An Evaluation of Piaget's Theory*, Brighton: Harvester 1981.

Gibson, J. J., *The Senses Considered as Perceptual Systems*, Boston: Houghton-Mifflin 1966.

Piaget, J., *The Origin of Intelligence in the Child*, London: Routledge and Kegan Paul 1953.

Sternberg, R. J., *Handbook of Human Intelligence*. Cambridge: Cambridge University Press 1982.

4

Enhancing intelligence

ROGER SCHANK and LAWRENCE BIRNBAUM

Researchers in artificial intelligence live in a theoretical world surrounded by a sea of anti-AI sentiment. Two of the strongest opposing
camps also oppose each other, thus lending to an already rather
treacherous intellectual environment an Orwellian aspect in which
convenient alliances are formed to allow for better attacks by any two
of the groups on the third. These alliances change quickly enough so
that it is not always easy for an outsider to tell which camp is which.

The views we are referring to are those typified by Noam Chomsky
and other linguists, by John Searle and other philosophers, and of
course, by us and other researchers in AI. These views conflict in a
number of ways. The debate often centres around the subject of what
machines can and cannot do. It sometimes hinges on certain ideas
about consciousness. It can also centre around the issue of what
makes man different from other animals. But it hardly ever directly
addresses what we will argue is one of the true sources of the conflict:
the issue of whether intelligence can be enhanced.

To identify the particular positions to which we are referring, we
will briefly describe them and give them names. The Searle position is
basically that no matter what a machine might do to imitate intelligent
behaviour, it needs consciousness actually to be intelligent. Searle is
unwilling to attribute consciousness, and hence true intelligence, to
a computer no matter what it does. However, he willingly attributes

consciousness to his dog, by virtue of its being happy to see him when he returns from a trip. This position we call the 'dog-consciousness position'.

Chomsky's view is quite different on this point. He makes no claims about whether dogs have consciousness, but he is quite clear that dogs and other animals do not have language. He argues that there is a kind of language organ that only humans have, and that it is the absence of this organ, more than simply the absence of laryngeal structures – or more to the point, of any other intellectual capacities – that accounts for the fact that chimps, for example, cannot speak. This position we call the 'language organ position'.

The position that we and other AI researchers hold is that one can build intelligent entities by analysing what intelligent behaviour consists of, determining the rules that govern that behaviour, and implementing those rules in a machine. Intelligence, under this view, is modifiable. Entities can become more intelligent if ways can be found to put more of the right stuff into them. This position we call the 'additive intelligence position'.

INTELLIGENCE AND EDUCATION

It is particularly important to address these three points of view on the mind at this time because of the fact that they have serious implications for our view of education. The problem of education is becoming a world crisis. At a time when there is more and more to know, more people seem to know less. And yet we seem to be incapable of providing means by which we might better educate people or by which they might better educate themselves. The underlying causes of our failures clearly lie in a myriad of difficult social, political, and economic factors. But among these and many other causes, there is a less obvious factor involved as well, and while it may be less pressing, it is something that is more directly under our control. Modern educational practice has been greatly influenced by trends in academic psychology. Behaviourism, for instance, inspired a large industry devoted to turning out educational products that put into practice what the theory preached. Of course, given the inevitable time lag in the popular dissemination of such trends, by the time this industry really hit its stride behaviourism was already in retreat as a theoretical framework for psychology. Unfortunately, by then the damage had been done.

We are faced today with educational doctrines structured by followers of positions not unlike the dog-consciousness position and the language organ position. These positions make certain claims about the nature of knowledge, its use, and its acquisition, and the results of their application to education have been curricula and teaching methods that don't work. The additive intelligence position, on the other hand, makes substantially different claims about knowledge and learning, which offer grounds for being far more optimistic about the prospect for enhancing intelligence. These claims form the cornerstone of our proposals here for changing education. To begin, however, we must first understand why the other two positions offer such a poor basis for educational theory and practice.

Chomsky makes little or no effort to discuss education or intelligence. The reason would seem to be clear enough. His perspective on language has direct implications for education, but they aren't the sort one would advertise. To the extent that Chomsky is offering his theories as paradigms for psychology – and he has certainly made no attempt to dissuade psychologists from using them in that way – then for any cognitive ability X the model is as follows. The ability to X is based in a mental organ specifically dedicated to X and nothing else. This ability has little if anything to do with any other cognitive abilities Y, Z, W, etc., each of which is based in a separate mental organ having little or no connection to either the X organ or any other. The structure and function of the mental organ for X (or Y or Z), and hence the ability to X (or Y or Z), is innate. In fact, the notion that we learn to X, Y, Z, etc., is largely an illusion; what happens is more akin to growth than learning. It may, of course, be necessary to tune or adapt the X organ and the others during maturation to perform effectively in the particular environment in which the organism lives. This tuning process consists in setting a relatively few simple parameters on the basis of observation, and takes place for the most part independently of any interaction with other mental organs dedicated to other cognitive abilities. Both the end point and developmental path of this tuning process are predetermined and they cannot be seriously affected either by environmental factors (except severe stimulus deprivation), or emotional or cognitive factors such as motivation.

From an educational perspective there is really only one redeeming implication of this model, and that is that it argues for providing all children with as rich an intellectual environment as we possibly can. In all other ways this is about as negative a view of the mind as could conceivably be imagined as far as the prospects for education are concerned. All we can do to educate people is to expose them to *X*, *Y*, and *Z*, and hope for the best. On this view there may be learning of a meagre sort – it would be more accurate to call it adaptation – but there can't be any teaching.

Chomsky argues that the innate genetic endowment governing language is 'universal', and this, along with impeccably correct political credentials, has saved him from the sort of harsh criticism that has been levelled at many others for holding similarly extreme views about the relative importance of innate versus environmental factors in intelligence. When it comes to other mental capacities, it has proven all too easy for other theorists to drop the assertion of universality. A number of years ago, for example, the *New York Times Magazine* ran an article on musical prodigies. The article included a sidebar in which a prominent educational psychologist offered the following account of the children's talents. Just as Chomsky postulated an innate mental 'language organ', so, he argued, it seemed likely that there was an innate mental 'musical cognition organ' as well. The explanation for the abilities of musical prodigies was then quite simple: their innate musical cognition organs are just better than yours and mine.

As unsatisfying as this explanation might seem, there really is no other way to account for individual differences in Chomsky's paradigm. On his view, the mind is a disconnected hodgepodge of special-purpose organs devoted to mathematics, music, and who knows, the latest fad in linguistic theory. But, obviously, some people are good at some of these things while others have different talents. Not only is there no obvious environmental account of individual differences in Chomsky's paradigm, but it is hard to see how environmental factors can play any serious role at all. We can do no more about our intellectual abilities than we can about our liver function.

This position can be, and has been, attacked in two ways. First, Chomsky's characterisations of both language use and language learning ignore much that is central to the picture. The abilities to learn and use language are in fact highly dependent upon other cognitive capacities. Jargon aphasics notwithstanding, you can't talk about what you don't understand; certainly you can't learn to talk without understanding. Second, to the extent that the picture Chomsky presents is correct, that is a reflection of the conventionalised aspects of language use, it cannot be considered a paradigm for cognition in general. No explanation can be offered for conventionalised behaviours because none is possible, except perhaps that some mutually agreed-upon choice is necessary (drive on the left or the right, write from the left or the right, say 'mama' to refer to your mother, and so on), and this happens to be it.

Most other natural cognitive tasks, including much of language use, are completely different in nature. Solving problems has its conventionalised aspects, but the ultimate correctness of a solution depends not on convention but on the world. We can look at a problem in the world, look at potential solutions, and see how and where they succeed or fail. Most importantly, we can see why they succeed or fail. Where such causal explanations are possible, a more cognitively active form of learning is possible. And so is a more active form of teaching.

John Searle's position on education is more complicated to figure out. Often his arguments have the cloak of 'obviousness' about them, in that he proves his points by claiming them to be obvious. Two of his well known obvious points are about the so-called Chinese room and his friendly dog. In the Chinese room argument, constructed partly in reaction to a visit to our laboratory in 1978, Searle asserts that a machine might be able to follow instructions and perform impressive feats of understanding, but that if a person were given those same instructions to follow, no one would claim that he understood what he was doing besides following rules. Thus, if the instructions were rules for speaking Chinese, and a human followed them, he would be able to speak Chinese without knowing how to speak Chinese or being able to understand Chinese.[1]

Searle's dog-consciousness argument extends this reasoning to point out that just as the instruction-follower could not have claimed to know Chinese in any important sense, no one could possibly claim that it was conscious. Following rules, no matter what behaviour might arise as a result, just doesn't look like consciousness to Searle. His example of a dog being happy to see his master, and thus appearing to be conscious of his master's relationship to him, is used to show that while dogs are conscious, computers never could be.

This paraphrase of his position may not be exactly what Searle intended, but it is not far from the mark. Our concern with this argument is not, however, with what it says about the potential for computer consciousness, although that is an interesting subject; rather, we are concerned here with what this argument says about intelligence and ultimately about education. The essence of the argument is that what you see is what you get. Dogs think, computers don't. No matter what you do to a machine it could never have any more consciousness, and therefore any more intelligence, than it has now (which is none, we suppose).

However much they may differ on the question of animal intelligence – whereas Chomsky believes that animals fundamentally differ

---

1 The crux of this argument turns on a deep confusion about computers and computer programs. Searle insists on identifying a locus of understanding in a single component of the computer system, in particular the rule-following component or central processing unit. Of course, no such single causal locus can be identified as the seat of a hypothetical computer system's capacity to understand language because no such locus can be identified as the seat of any computer system's ability to perform any task in executing any program, whether its purpose is to understand language or to calculate the payroll.

from humans, Searle believes they are fundamentally the same – Chomsky and Searle both take what might be termed an essentialist position on human intelligence, and their views have curiously similar implications for human education. For if Searle is correct, then intelligence is not augmentable. There is nothing that you can do to make a thing more conscious, and hence more intelligent, than it already is. Giving rules or facts or experiences to computers will not make them intelligent because it will not make them conscious. Adding more rules, facts, or experiences will not make them more intelligent because the essence of intelligence, in this view, does not lie in rules, facts, or experience. In fact these things have nothing to do with intelligence because they have nothing to do with consciousness. It is something else entirely that is at the root of intelligence.

But if adding facts, rules, behaviours, or experiences cannot make a difference, if we cannot find out what stuff to add to an entity to endow it with more intelligence, if this enterprise is hopeless as Searle maintains, then education would seem to be hopeless as well. A dog doesn't learn to be conscious after all. It has exactly as much of the stuff as it has and that's all.

The position we adopt here is much closer to Charles Darwin's than to Searle's or Chomsky's. We maintain, as did Darwin, that there isn't much difference between human intelligence and the intelligence of higher animals. Further, we maintain that intelligence is enhanceable, and that what AI has discovered about how to enhance the intelligence of machines tells us a great deal about how to make people smarter, and thus how to change education. In contrast, the views of the mind and of intelligence that both Chomsky and Searle adhere to seem to us inherently anti-education because they both entail, as a central tenet, that what intelligence exists in an entity is completely and inherently unalterable in any positive way. We can only speculate as to why anyone would take satisfaction in such a view of human beings and human potential.

ASPECTS OF INTELLIGENCE

How can we hope to answer questions about whether machines can be intelligent without knowing what constitutes intelligence? Yet, rather surprisingly, such issues are rarely part of the intellectual debate about the possibilities for AI. Even more surprisingly, they are often not part of the debate about human capabilities either. When we hear linguists talk of innate human capabilities and 'language organs', what we signally fail to hear about is the employment of these capabilities. The human ability to create sentences that have never before been created is certainly a wonderful and complex process. But does it make sense to study this process independently of its function? Intelligent entities have the intention to communicate, and they use a variety of intellectual abilities in order to achieve this goal. Does it make sense to study language as a psychological phenomenon independently of memory-retrieval abilities, storytelling abilities, inferential abilities, or abilities to reason about the expected impact of new information upon a hearer?

Despite these questions, through devices such as the competence-performance distinction many scholars have concluded that scientists need only be concerned with 'competencies' addressed in a 'mathematically rigorous' fashion. This stratagem, and its general acceptance by many linguists, psychologists and philosophers, has had disastrous consequences for research in these fields. This is a shame, but what is really much worse has been the consequences for our notions of what belongs in a student's education. To cite one example here, consider the idea of diagramming sentences. Grammar, and in particular, diagramming sentences, is a part of every school child's education. It is true that the practice of this particular skill long predates Chomsky's theories; nevertheless sentence diagramming, and grammar study generally, enjoyed a considerable resurgence after his work became widely known. Furthermore, the narrowly mechanistic philosophy of education from which it follows that schoolchildren ought to be diagramming sentences bears a considerable resemblance to the philosophy of research underlyng Chomsky's own work in linguistics.

To see what we mean, consider the five aspects of intelligent behaviour cited above: language, memory-retrieval, storytelling, inference, and expectation. All of these capabilities are equally important aspects of the process of sentence creation, which is itself only one aspect of intelligence. Yet, while linguists invent formal rules purporting to describe the 'mathematics' of language, and school children are encouraged to learn aspects of these rules, the other subjects remain virtually ignored both in scholarship and in school. Why might this be?

Certainly, it would be hard to argue that we can speak or write independently of memory. Chomsky rules out memory considerations as irrelevant for the study of language, treating memory as something whose main job it is to retain sentence structure plans. Yet can we speak of what we don't know? Can we write about events we do not remember or imagine? The process of finding memories and manipulating them to suit our needs for the purposes of communication is just as important a part of the communication process as language is *per se*. In fact, there is a great deal of reason to consider it to be a far more important part of the communication process, since, whether or not dogs have consciousness or chimps have language, it is perfectly clear that both have memory retrieval capabilities.

Memory retrieval capabilities are not studied in school: there is no examination of the process, no diagramming of rules, and no discussion of the issue. Why? One might imagine that the answer is that not a great deal is known about them, and to some extent this is true; but the real reason is less obvious, and lies in the unarticulated biases underlying much work in cognitive science.

AI AND THE BIAS AGAINST CONTENT

The special contribution of artificial intelligence to the study of the mind is often supposed to lie in its stress on the creation of process models of mental abilities. In fact, however, this is only a small part of the story. What makes AI unique among the cognitive sciences is its stress on functionality, on the idea that the chief constraints on the mind, and hence on models of the mind, must arise from the need to perform realistic cognitive tasks. From the perspective of AI, intellectual capacities cannot adequately be studied without considering the purposes to which they are put by the organism which possesses them. This is exactly what Chomsky denies explicitly, and Searle implicitly. It is also what most differentiates AI from psychology and linguistics.

The difference can most clearly be seen when we consider the 'process models' that are often proposed by researchers in these allied fields, and may indeed in some cases even be implemented on computers. However sophisticated such models may be in their own way, they are almost always completely inadequate from an AI perspective. Why? Quite simply, it is because for the most part they are models of mental phenomena and not of mental abilities. In the hands of linguists and psychologists, computational modelling is simply a novel variant on what is termed, in those fields at least, 'mathematical modelling': it is basically curve fitting by another means. The point of such models is simply to generate the phenomenon of interest. It doesn't matter in the least whether the resulting programs are capable of performing an intelligent task of any sort. The question, 'What is this program for?' makes no sense when applied to these models. From the perspective of computer science, however, this is the very first question that must always be asked, and answered, about any program.

The result of AI's stress on functionality is to direct our attention in completely different directions. When a linguist or psychologist looks at a topic in cognitive science, the question of interest is often, 'What makes this different from other cognitive phenomena?' It is more or less assumed that the essence of the phenomenon lies in what makes it unique, and that concentrating on this issue is the key to understanding the nature of the phenomenon. From a taxonomic point of view, this is of course a sensible notion. From a functional point of view, however, it is much less sensible. For if our goal is to build a model that is capable of carrying out some intelligent task, the issue isn't 'How is this task different from all other tasks?' The issue is, 'What is the bottleneck problem here? What makes this task difficult?' Often, we find that the underlying functional issues are the same, even if the tasks look superficially quite different.

Consider, for example, the ability to create novel sentences that we discussed in the previous section. The need to account for this capacity in some way forms the cornerstone of generative linguistics. And yet it would be very surprising indeed if the generative nature of language were an isolated phenomenon in mental life. After all, if we never had anything new to say then a simple table would in fact suffice to represent the mapping between thought and language. From a functional perspective, therefore, the generative nature of language is necessitated by, indeed is merely a reflection of, the generative nature of thought itself. It is not our ability to create new sentences that needs explanation: it is our ability to create new thoughts. Whether the task is learning or language understanding, the key problem is how to represent new ideas in terms of old ones.

The functional perspective outlined above is the key to understanding the central insight of artificial intelligence. It is a simple point really. What makes someone intelligent is what he knows. What is needed to make intelligent computers is to endow them with knowledge. This was not the discovery that we set out to make. Many of the founders of AI came out of mathematics and physics, and were trained in a scientific aesthetic which put primary emphasis on the search for powerful yet extremely general principles. What they aspired to discover was something like the law of universal gravitation for the mind. In retrospect it would have been far better for everyone concerned if the founders of AI had been trained in biology instead!

What drove AI researchers to abandon the search for a few simple principles was hard experience. Every project to build an intelligent program, no matter how limited in scope, ended up confronting the same bottleneck: to get the computer to do something interesting, it had to know a great deal about what it was trying to do. Gradually it dawned on people that if knowledge was the functional bottleneck in building intelligent machines, then it followed that knowledge was the central factor in intelligence generally. However obvious this discovery might seem in retrospect, most of cognitive science (indeed, much of AI itself) still seems determined to avoid its implications. One reason for this resistance is that the prescription for AI that follows from this discovery doesn't look like science to a lot of people: it just looks like writing down what your grandmother knows. From this you can make a science?

The attack on the notion that we might build a psychology around mental content comes on two fronts. On the one hand, we have Chomsky arguing that, for example, the commonsense theory of human intentionality necessary to understand the behaviour of people around us isn't in the domain of psychology at all; it should be left to literature. Roughly speaking, he endorses the computational view of the mind but rejects as pointless and unscientific efforts to bring mental content into the computational realm. On the other hand, while Searle clearly endorses the notion that mental content is the essence of intelligence, he denies that any computational account of such content is possible. This is the core of his argument against AI. Thus he shares Chomsky's view that efforts to bring mental content into the computational realm are pointless, but for an entirely different reason: not because mental content is 'unscientific' but because its essence would be completely missed by such an enterprise. In sum, AI is condemned both for attempting to do semantics at all, and for doing such a poor job of it.

DARWIN AND ARTIFICIAL INTELLIGENCE

The issues of intelligence and intelligent behaviour were a central scientific concern of Darwin's. In particular, he found it necessary to state a position on the intelligence of animals as part of his arguments about the descent of man from other species. If animals have no intelligence at all, it is much more difficult to believe that intelligent man could have naturally descended from them. Thus Darwin was led to claim that 'there is no fundamental difference between man and the higher mammals in their mental faculties'. It seems obvious that many cognitive scientists working today would sharply disagree with this assertion, and for that reason alone it is interesting to look at some of Darwin's arguments. Further, it seems to us that Darwin would find AI to be a field whose premises are consistent with his own observations.

It is this latter point with which we will start. A central aspect of intelligence is the need to generate and answer questions. No entity can learn without generating for itself the need to know. When Darwin writes that 'all animals feel Wonder and many exhibit Curiosity', he is putting this forward as the key to the 'intellectual faculties' of animals. When he notes that 'hardly any faculty is more important for the intellectual progress of man than Attention', he is making the point that in order to learn, one must first know where to look, and then arguing that animals can focus their attention in just this way. Darwin tells a story to illustrate these issues, about a baboon that adopted a kitten and, upon being scratched by the kitten, located the kitten's claws and bit them off.

Darwin used this story to point out that animal behaviour cannot entirely be attributed to instinct. What kind of instinct would this have been for the baboon? It seems clear enough that the baboon figured out what to do in a novel situation. This behaviour exemplifies the same understanding cycle that forms the cornerstone of human intelligence: a cycle of expectation failure, followed by curiosity, followed by an explanation derived from a prior similar experience by virtue of memory recall, followed finally by a generalisation that will itself fail at some later time. A computer also can act in this way to the extent that we understand well enough how to represent and arrange these sorts of actions in its memory. If Darwin would attribute intelligence to the baboon, he would attribute it to the computer as well.

What arguments might he marshal against such a premise? One standard objection might have to do with the fact that the baboon was in no way programmed to behave in this way; its behaviour was entirely original. The computer, so the argument goes, would have to be programmed in order to exhibit such behaviour. This sort of argument raises two general questions. First, is it possible to program a computer with general knowledge that would enable it to solve the baboon's problem with no specific prescription for biting off kittens' claws? Second, is it possible that the baboon had no specific prescription for biting off kittens' claws, but was simply programmed in a manner similar to the way a machine might be programmed?

Darwin refers to an elephant in the zoo who 'blows through his trunk on the ground beyond [an object] so that the current reflected on all sides may drive the object within reach. These actions of the elephant . . . can hardly be attributed to instinct or inherited habit, as they would be of little use . . . in a state of nature'. Darwin is arguing, in essence, that figuring something out for yourself is what intelligence is all about. If elephants and baboons are intelligent to the extent that they can figure things out for themselves, then the question for the creation of intelligent entities is whether we can enable them to figure things out for themselves, not whether or not they are already endowed with this ability.

Thus, it turns out that the fields of education and AI are asking the same question. For Darwin, it seems that any species that figures things out for itself qualifies as intelligent, and he would no doubt be unconcerned if it were discovered that sea slugs also were able to reason. It is for this reason that we can assume that Darwin would be prepared to view computers as intelligent entities, whereas Searle and Chomsky insist on denying this idea.

TEACHING ARCHITECTURES

The notion that education and AI are concerned with fundamentally the same question has played a key role in leading us to consider the question of how computers can be employed in improving education. The application of computers to education has been disappointing to date. It has almost always entailed a particular teaching architecture, something we can call the page-turning architecture, which has been left implicit in the design of educational software and is responsible for the lack of excitement or educational value of most software on the market today. Page-turning architecture involves putting up a screen of information and asking the student either to indicate when he wants the next page of information or to answer a question that will cause him to get another page of information. Clearly there is a problem here. The model assumes that information given to a student is absorbed by the student, like a sponge, regardless of how interested the student is in the material. The student is then tested to assess how much has been absorbed. And this, by and large, is our current educational model. Certainly we can do better.

We present here four teaching architectures. These assume that learning involves more than reading and answering, or that if reading and answering are involved, then that is precisely what the student wants to do at the moment that the computer engages in that form of interaction. Going beyond page-turning architecture requires a new approach to software. For one thing, it entails enabling computers to store and retrieve far more information than has been typical so far. It also entails representing that information so that it is accessible at the right time for the right reason. In other words, to make machines better instructors, we must give them better information, in an accessible form, and we must design teaching architectures that make use of such information.

The four architectures presented here are inspired by what we know of how people learn and the information that we believe can be imparted to machines that will enable them to be better instructors. These architectures, then, are artificial intelligence architectures for the design of teaching systems.

Case-based teaching architecture The first architecture we discuss depends upon two ideas. The first is that experts are repositories of cases. When a doctor, lawyer, or auto mechanic is doing his job, for example, he is likely to rely upon his knowledge of previous cases to help him in current decision-making. He might recognise a situation he is working on as quite like one he has previously encountered. His task is then to draw upon the similarities and differences in the case he has been reminded of, to help him better understand the new case. This well-recognised approach to understanding and problem-solving is called case-based reasoning. It is clear that much reasoning by experts relies upon a significant case base that can be utilised when necessary.

The second key idea in this architecture is that good teachers are good storytellers. For a student to benefit from a case that the teacher knows, three conditions must be met. First, the student has to be ready to hear about the case, in other words he has to be in a position to use the information that is contained in the case for something he is working on or is interested in. He also has to know that he is in a position to use the information in the case. And, lastly, the case has to be told in such a way as to capture and hold his interest. In short, when he is ready to listen to a good story, he is in a position to learn from that story.

The premise of software that exploited a case-based teaching architecture would be to place a student in a situation that the student found inherently interesting. Such a situation might involve, for example, having a student attempt to build or design something on the computer. This basically creative task should be one that is inherently appealing to a student so that there would be no problem in getting him interested in performing it. The task should be complex enough that all that the student might need to know is not available to him. The task of the program would be to teach the student what he needs to know, or what he might consider while doing his task, at precisely the points in the task in which he becomes interested in knowing it. This information should be presented in the form of stories.

Inherent in the case-based teaching architecture is the idea that learning takes place on a need-to-know basis. The case base must be indexed in such a way as to relate to the situations that are encountered in performing the task. The juxtaposition of a case base and a situation that indexes the case base is the essence of case-based teaching. This kind of teaching corresponds to how learning actually takes place: learning occurs when students really want to know what someone has to teach because knowing that information will be useful in a way that is readily apparent to the student. In this context, the student will be able to understand the information, and will be able to index it in his own memory in a way that will facilitate its retrieval in appropriate situations in the future.

Incidental learning architecture Not everything is fun to learn. In fact, some things are terribly boring to learn. And when some curriculum inventor decides that such things must be mastered, you can be sure that he will try to teach them as a list to be memorised or a set of items to be tested upon. But people do habitually learn a variety of information that is quite dull without being completely bored by it. They do this by not setting out to learn that information at all.

Much of what we know we have learned in passing. We have accumulated knowledge simply by the act of living and by being in situations in which that knowledge came up and was of use. Our knowledge comes from experience and is thus scattered around memory, stored with those experiences. We may find it difficult to retrieve a set of facts about Iowa, for example, when asked to produce them, but we may find that information when we need it, when it comes up in a natural way. Much of what is taught in school involves information presented in a manner quite different from the way it is naturally used. We have lessons in which we expect students to learn lists of items or sets of facts. The method we use is to tell them these facts and hope that they remember them. But we can much more reasonably expect students to remember facts that they gathered for some use, for some real purpose in which they were interested. The task, then, is to design software that will present students with situations that are inherently interesting. Then we can use those situations to teach some particular items of information by causing the interesting situations to present themselves only if the desired material is learned. It is very important to keep in mind, however, that the student's interest is strongly related to the extent to which the material to be learned and the reward material really are intertwined in a natural way.

The architecture here, then, is the creation of tasks that are inherently dull but whose end result is inherently interesting. There are likely to be many methods of accomplishing this, and hence many sub-architectures involved. Each method would serve as a standard software tool into which relevant material could be put to create new learning situations. The key to exploiting incidental learning is to find things that are inherently fun to do on a computer. This could be any good video game, for example. The next part is trickier. What the student naturally wants to learn in the video game ought to be worth learning, such as content-based tasks, for which one needs to know real information in order to accomplish one's goal on the computer. This will work very well if there is a natural correlation between the content-based tasks and what is inherently fun.

Directed exploration of video database connections It is quite usual to teach by giving assignments to students. Students are asked to write reports by going to the library and researching a topic. The concept here is a good one. Let students discover what they might find interesting and then have them organise what they find into a coherent report. This method of instruction has its flaws, but it can be quite effective if employed correctly. Even more effective, however, if they could be made easily explorable, would be to let students explore large video databases.

A tremendous amount of information is available on video. There exists in the archives of television networks, for example, a vast amount of footage of the important news events of the world over the last thirty to forty years, of important leaders in every field being interviewed, of studies of animals in the wild, of instructional material, and so on. Imagine if that material were available to any student who wished to view it. While print authors can often analyse a subject very effectively, nothing compares to seeing it for yourself. To make computers really useful information providers, we must create learning environments in which each user can follow his or her own interests. We must make use of the members of our society who have important things to say, who have had important experiences that decision-makers ought to be aware of. It is important to make the breadth and depth of knowledge in our society available to everyone who might need it.

The problem here is one of organisation. It really doesn't help anybody to make thousands of hours of video available without an organisational scheme that allows a user to find what he wants almost instantly. Further, this information is of no use if the system cannot tell the user about information it has available that he would not have known about at the moment when he might have become interested in that information. Learning depends upon good information being available at the time one is ready to hear it. The major technical problem in constructing a properly organised video library is therefore content-based indexing. In order to tell a story at the proper time, either the computer must have assessed the situation well enough to have understood that a given user needs to be told a certain story, or the user must naturally 'run into' the proper story in the natural course of what he is doing.

There are many possible questions that an individual might ask after hearing someone say something or after reading something. This architecture should make it possible to ask these questions and have the answer available as the very next piece of information presented. Video databases would be valuable things if they were indexed in such a way as to present information organised by the content of that information, always pointing the way towards other available information that is related in content. In any scheme of this sort, any one video clip would have multiple indexes, since it could relate to many possible points.

The rewards of such an undertaking, if properly executed, would be tremendous, not only for helping to interest students in finding out about the world, but as an archive of past history useful for all kinds of different purposes.

Simulation-based learning by doing There is really only one way to learn to perform an activity, and that is to do it. If you want to learn to throw a football, drive a car, build a mousetrap, design a building, teach students, cook three-star cuisine, be a management consultant, you must simply learn to do it by doing it. However, for many of these sorts of tasks, rather than allowing students to learn by doing, we create courses of instruction that tell students about the task in a theoretical way, without concentrating on the performance of the task. To put this another way, when apprenticeship cannot be easily or affordably implemented, we try lecturing.

The teaching architecture implied by this is to change every possible skill into a learn-by-doing situation. To do this, one must create simulations that effectively mimic the real-life situation well enough to prepare the student for such situations without having actually been in them. The flight simulator has been effectively employed in the training of pilots because there was money available to build such simulators, and because the risk of injury in real-life learning-by-doing was high.

Simulations of all kinds can be built. What is required is to understand the situation to be simulated well enough that the simulations will be accurate portrayals along the dimensions that matter in performing the task. This can mean, in the case of simulations of people-to-people interactions, creating complex models of human institutions and human planning and emotional behaviour. These simulations would have to be built to have roles available that the student could play. Thus, to learn to be a loan officer in a bank, he would play the role of loan officer in the simulation, trying out new situations that would cause him to have simulated experiences analogous to those he might encounter in the real world.

There are at least four basic components to such learn-by-doing environments: the simulator itself; a teaching program that helps the student through the simulator and is able to discuss issues with him; a language understanding program that can comprehend what a student might ask the teacher; and a storytelling program that would tell stories representing the experience of experts in situations encountered by the student in the simulation, and that would be activated by the teacher when appropriate. This teaching architecture is critical when the subject matter to be learned is one that is really experiential at heart. Much of natural learning is the accumulation of experience. Schools have trouble allowing children to learn from experience both because their individual experiences are so different and because the classroom situation does not allow for much actual experience to occur. Nevertheless, experience is the best teacher.

WHERE'S THE AI?

At the Institute for the Learning Sciences, we have begun to incorporate the four teaching architectures described above, among others, in building high-quality educational software for use in business and elementary and secondary schools. We have created quite a few prototypes that go beyond the kind of things that have traditionally constituted educational software and that change the nature of how students interact with computers. These prototypes have been very successful, but, nevertheless, we have been plagued by a curious kind of question when we show them to certain audiences. The question we keep hearing is, 'Where's the AI?'

What assumptions about AI are inherent in this question? It seems to assume at least one of the following four things:

- AI means 'magic bullets'
- AI means inference engines
- AI means getting a machine to do something you didn't think a machine could do (the 'gee whiz' view)
- AI means having a machine learn

The magic bullet view of AI is this: Intelligence is actually very difficult to put into a machine because it is so very knowledge-dependent. Since the knowledge acquisition process is very complex, one way to address it is to finesse it. Let the machine be very efficient computationally so that it can connect things to each other without having to explicitly represent anything. In this way, the intelligence will come for free as a by-product of unanticipated connections that the machine makes.

The inference engine view of AI was brought forth when the expert systems world began to explode. AI people were expert at finding out what an expert knew and instantiating that knowledge in rules that a machine could follow. While such expert systems could in fact make some interesting decisions, business types who were called upon to evaluate the potential of such systems probably asked 'Where's the AI?' There had to be some specific thing that could be labeled 'AI' (as with Searle's similar confusion, above pp. 76–7), and that thing came to be known as an inference engine. Of course inference is an important part of understanding, but to label the inference engine as the AI was both misleading and irrelevant. Business people who assumed that AI and expert systems were identical began to expect inference engines in anything they saw on a computer that 'had AI in it'.

The next view is the 'gee whiz' view. This maintains that if no machine ever did a particular task before it must be AI. The best examples of what happens with this conception of AI are chess-playing programs. Are these AI? Today most people would say that they are not. Years ago they were. What happened? The answer is that they worked. They were AI as long as it was unclear how to make them work. When all the engineering was done, and they worked well enough to be used by real people, they ceased to be seen as AI. Why? First, 'gee whiz' lasts only so long. Second, people confuse getting a machine to do something intelligent with getting it to be a model of human intelligence – and surely these programs aren't very intelligent in any deep sense. Third, the bulk of the work required to transform an AI prototype into an unbreakable computer program looks a lot more like software engineering than it does like AI to the programmers who are working on the system; so it doesn't feel like you are doing AI even when you are.

The fourth view of AI is the one that we have espoused here. It says that intelligence entails learning; intelligence means getting better over time. Real AI means a machine that learns. The problem with this definition is that according to it, no one has actually done any AI at all, although some researchers have made some interesting attempts. This means that 'Where's the AI?' is a tough question to answer.

AI entails massive software engineering. To make AI – real AI – programs that do something someone might want, one must do a great deal of work that doesn't look like AI. To rephrase Thomas Edison, 'AI is 1 per cent inspiration and 99 per cent perspiration'. We will never build any real AI unless we are willing to make the tremendously complex effort involved in making sophisticated software work. But this doesn't answer the question. Sponsors can still honestly inquire about where the AI is, even if it is in only 1 per cent of the completed work. The answer to 'Where's the AI?' is then, 'It's in the size'.

Prior to 1982, AI had a real definition, and it was the 'gee whiz' definition given above. But underlying even that definition was the issue of 'scale-up'. AI people always had agreed among themselves that this was the true differentiation of what was AI from what was not AI. This measure of AI was one of those things that was so clearly a defining characteristic of the field that there was no need to actually define it on paper. An AI program is one that is based upon a theory that is likely to scale-up. The point here is that the AI is in the size, or at least the potential size. The curious thing is that when the size gets big enough, this all ceases to matter. Again, we merely need to look at AI work in chess. Getting a machine to do something that only smart people can do seemed a good area for AI to work on. Now, many years later, you can buy a chess-playing program in a toy store and no one claims that it is an AI program. What happened?

The task in AI was to create intelligent machines, which meant having them exhibit intelligent behaviour. The problem was, and is, that exactly what constitutes intelligent behaviour is not exactly agreed upon. Using the scale-up measure, the idea in looking at a chess program would have been to ask how its solution to the chess problem scaled up. We need to know if the solution to making such programs work tells us anything at all about human behaviour. We would also want to know if it tells us something that we could use in any program that did something similar to, but more general than, playing chess. In fact, the original motivation to work on chess in AI was bound up with the idea of a general problem-solver. The difficulty is that what was learned from that work was that people are really specific problem-solvers more than they are general problem-solvers, and that the real generalisations to be found were in how knowledge is to be represented and applied in specific situations. 'Brute-force' chess programs, which performed high-speed search through a table of possible pre-stored moves, shed no light on that issue at all, and thus are usually deemed not to be AI.

The fact that a program does not scale-up does not necessarily disqualify it from being AI. The ideas in a program may be AI ideas without necessarily being correct AI ideas. What then, does it mean for a program to have AI ideas within it? This is, after all, a key question in our search to find the location of the AI.

When we talk about 'scale-up', we are, of course, talking about working on more than a few examples. It is critical, if AI is to mean applications of AI ideas rather than simply the creation of those ideas, that size issues be attacked seriously. No one needs a program that does five examples. This worked in 1970 because AI was new and glossy. It will not work any longer. AI has to dive headlong into size issues.

The truth is that size is at the core of human intelligence. In a recent book entitled *Tell Me a Story*, the first author argued that people are really best seen as storytelling machines, ready to tell you their favorite story at a moment's notice. People rarely say anything that is very new or something they have never said before. Looked at in this way, conversation is dependent on the art of indexing, of finding the right thing to say at the right time. This is a pretty trivial problem for someone who only has three stories to tell. Many a grandfather has survived many a conversation on the same few stories. But when the numbers get into the thousands, one has to be clever about finding, and finding quickly, germane stories to tell. We cannot even begin to attack this problem until the numbers are large enough. In order to get machines to be intelligent they must be able to access and modify a tremendously large knowledge base. There is no intelligence without real, and changeable, knowledge.

And this thought, of course, brings us back to the original question.

If a system is small, can there be AI in it? How big is big enough to declare a system an AI system? It is fairly clear that while this is an important question, it is rather difficult to answer. This confronts us once again with the issue implicit in all this: the concept of an AI idea. When we said earlier that certain programs have AI ideas within them, we implied that even programs that did not work very well, and never did scale-up, were AI programs. What could this mean?

AI is about the representation of knowledge. Even a small functioning computer program that someone actually wanted could be an AI program if it was based on AI ideas. If issues of representation of knowledge were addressed in some coherent fashion within a given program, AI people could claim that it was an AI program. But the key point is that the question 'Where's the AI?' is never asked by AI people; it is asked by others who are viewing a program that has been created in an AI lab. To these people, crucially, it simply shouldn't matter. The answer to the question, when a program does a job that someone wanted to do, is that the AI was in the thought processes of the program's designers and is represented in some way in the programmer's code. However, if the reason that they wanted this program in some way depends upon the answer to this question – that is, if they wanted 'AI' in the program they were sponsoring – they are likely to be rather disappointed.

AI depends upon computers that have real knowledge in them. This means that the crux of AI is in the represention of that knowledge, the content-based indexing of that knowledge, and the adaptation and modification of that knowledge in the light of experience with its use. In our view, therefore, case-based reasoning is a much more promising area than expert systems ever were, and within the area of case-based reasoning, the most useful and important – and maybe even somewhat easier – area to work in is case-based teaching. Building real, large, case bases, and then using them as a means by which users of a system can learn, is a problem with enormous import for both AI and the users of such systems that we can attack now.

WHAT IS AND IS NOT TEACHABLE AND WHY
IT MATTERS TO KNOW THE DIFFERENCE

It is fairly obvious that some aspects of what is commonly referred to as intelligent behaviour are inherently unteachable and that others are teachable. Which are which is somewhat elusive, partly because there is no clear agreement on what constitutes intelligent behaviour in the first place. Let us therefore look at some aspects of intelligence that we have learned are critical to intelligent behaviour, in the sense that no machine could exhibit much intelligence without these abilities. Then we discuss which of these abilities are likely to be human-specific. Finally, we speculate about the teachability of these abilities.

Let us start simply, with a discussion of scripts. A script is a memory structure that accounts for expected behaviours in mundane and routinised environments. Scripts often define behaviour in complex human environments involving many players. Scripts tell us who is doing what, why they are doing it, and what we should do about it. The paradigmatic script (restaurants) tells us what the waiter will do, what we are expected to do, and in what order these actions should occur. Given the importance of scripts in human behaviour, the key questions concerning them are: Do machines need scripts in order to exhibit intelligent behaviour? Are humans born with them? Do animals have them? Do either humans or animals learn them? Can they be taught?

From the point of view of AI, the answers to these questions are straightforward. Machines come *tabula rasa* so there are no scripts. However, scripts can be programmed into them, allowing them to read and answer questions about texts that depict script-based behaviour as well as enabling them to participate in scripts. Although there is much in human behaviour that is not script-based, there is much that is. Endowing machines with scripts enables them to mimic that behaviour.

There is no evidence that humans are born with scripts although there is a good deal of evidence to suggest that they are born with a predisposition to acquire them. Small children seem to create scripts at every turn, expecting events that have followed a particular sequence to repeat that sequence the next time. Animals exhibit the same behaviour. They too establish sequences at an early age and act as if they expect these sequences to repeat. While they do not use scripts for language understanding, they nevertheless seem to be capable of acquiring them and acting in accord with them. Such behaviour has frequently been stigmatised as 'just association' by people from William James on down, but it is difficult to understand the 'just' here. Animals and people learn to associate one event with another, but their recording of these associations and reliance upon them for future processing is one of the hallmarks of intelligence. If a script is not built in, if it is learned, and if this learning is critical for intelligence, then it follows that intelligence is indeed enhanceable.

One script that is clearly not built in is the airplane script.[2] After many flights, we understand about such things as seat belts, baggage X-rays, boarding passes, tray tables, overhead luggage compartments, and so on. Our model of air travel liberates us from the burden of thinking hard about what to do on a plane when we are hungry, as well as enabling us to understand sentences such as 'I couldn't sleep because of the leg room', which would otherwise be difficult if not impossible to comprehend. If a dog learned about behaviour on airplanes we would have to say that it knew some aspect of the airplane script. And, in fact, dogs do learn such scripts, constantly finding certain places and ways to behave that are at first novel and then repetitious. Scripts, then, are learnable. More importantly, they are teachable. Children can learn about new behaviours in new situations and utilise that new knowledge in a variety of ways.

This question of what is and is not teachable is significant for two quite different reasons. From a scientific point of view, the question

2 Although Fodor, for example, has gone so far as to argue that concepts like 'telephone' must be innate, on the grounds that necessary and sufficient conditions for being a telephone cannot even be written down by philosophers, let alone learned by ordinary people. This he apparently finds more plausible than the alternative that concepts are not defined in terms of necessary and sufficient conditions.

of learnability is quite often the hallmark of the underlying assumptions of a psychological theory. The generative linguists came to the somewhat remarkable conclusion in the 1970s that the grammars they proposed were not learnable. From this they drew the conclusion that these grammars were innate. They might, of course, have considered the possibility that their theory was wrong, but they failed to do so. That they failed to do so indicates a very strong (if unstated) position on education, namely, that much intelligent behaviour is innate and that education that is intended to affect core abilities is wasted. It is important to understand that this assumption is very prevalent in today's educational establishment, and while these ideas are not necessarily attributable to Chomsky, they are not unrelated either.

For example, one of the prime distinctions in this regard in American society is attributable to the Educational Testing Service (ETS). ETS administers two very important exams that in part determine the life of every prospective college student in the United States, the Scholastic Aptitude Test (SAT) and the individual Achievement Test. The premise of the SAT is that one cannot study for it. It is supposed to test basic aptitudes. The fact that an industry has grown around tutoring for the SAT to do exactly that which ETS claims is impossible to do, does not sway the ETS from its position. The Achievement Tests, on the other hand, are supposed to measure what one has learned and hence what is learnable. Thus, ETS has operationally defined the difference between core abilities or basic aptitudes on the one hand, and acquired knowledge on the other. Where on this spectrum would something like the Pythagorean theorem go, would you suppose? It is a little surprising to discover that a question that involves knowing the formula derived from this theorem appears on virtually every SAT test. It would seem odd (although not perhaps to Plato) to claim that this theorem is innate, but ETS does claim that one cannot be tutored for the basic aptitudes measured by the SAT. All this may leave us somewhat confused.

This same confusion manifests itself in the attempt to understand Chomsky's theories. Chomsky explicitly rejected the notion that in a sentence such as 'John likes books', the idea of reading ought some-

ROGER SCHANK and LAWRENCE BIRNBAUM

how to be part of the deep structure; or that in a sentence like 'Mary hit John because she was angry', the concept of revenge ought somehow to be present. To admit this possibility would (and eventually did) lead to the requirement that a linguistic theory would have to encompass more than purely linguistic phenomena. Or, to put this another way, if Chomskian theory began to be about human experiences and inferences based on memory, then it would cease to be purely about linguistic competencies and thus about what specifically was held to be innate.

The innateness hypothesis, then, is not at all random happenstance. The only way one could possibly postulate a theory of language independent of real-world knowledge was to claim innateness for the basic linguistic competence. If this claim were not made, one would have to admit that different people, by virtue of their idiosyncratic real-world knowledge, understood language in highly idiosyncratic ways that depended upon the particular knowledge they happened to have. Such an 'unscientific' hypothesis would have been too much for the Chomskians who want desperately to create universal theories unsullied by 'grandmother knowledge'.

But the price of all this scientific neatness is too great, for it presupposes an answer to the question of whether knowledge is something one acquires after certain basic intellectual competencies have been determined, or whether knowledge and such competencies are inherently intertwined. If knowledge and its proper use are not an important part of the basic competence of intelligence, then nothing of great consequence is altered by acquiring more knowledge. All that happens is that people who are at certain basic intellectual levels stay at those levels but happen to know more. If, however, knowledge is an integral part of the process of thinking, then its acquisition is of critical importance in attaining new intellectual levels. The claim in this case is that in order to think well, to be intelligent, one must know a lot. Chomsky's theory denies this claim. AI theory vigorously supports it. Our whole concept of what ought to go on in school depends on this concept of the possibility of the enhancement of intelligence.

To put this another way, the traditional view of intelligence – and modern linguistic theory is completely in line with this tradition – is that knowledge plays a passive rather than an active role. It is simply grist for the mill. Serious psychological effort must, then, be directed at the mill, not the grist, for it is the mill that captures the universal, necessary essence of intelligence. Moreover, the mill has clear mechanical structure to be studied, while the grist just assumes whatever form you put it in. In sum, knowledge is simply too contingent, too atomised, too undifferentiated, to form the core of any serious psychological theory. This 'minimalist mentalism', it appears, reflects the very same prejudices about what constitutes science that led to behaviourism.

There are traces of this attitude within AI as well, it must be admitted. As we noted above, the expert systems community, which originally started out by fervently proclaiming the importance of knowledge (for example, coining the term 'knowledge-based systems'), ended up by producing inference engines instead. What had originally been seen as the key to effective performance became something to be put in by essentially untutored 'knowledge engineers' after AI specialists finished the hard technical work of designing and implementing the overall system architecture. Similarly, work in knowledge representation often degenerated into the design of utility packages or inference 'services' designed to perform some limited class of inferences (such as inheritance inferences), and to a deep preoccupation with the syntactic properties of representation systems. If asked why they were moving in these directions, so profoundly counter to the spirit that originally motivated these enterprises and even to the names they called themselves, most of the researchers involved would undoubtedly say that they were forced to do so by considerations of generality – just what the linguist would probably say. In other words, they shared the belief that from what your grandmother knew you couldn't make a science.

One could, of course, take an entirely different approach to knowledge. Individual facts, considered individually, will necessarily look atomised and entirely contingent. But facts do not arise individually in the world, nor are they represented that way in the mind. Facts are coherent; they make up and are derived from systematic theories of the world. We can neither understand, remember, nor utilise isolated facts that we cannot relate in a systematic way to a larger body of knowledge. It is therefore these larger bodies of knowledge that must form the core of our study of intelligence. Your grandmother doesn't just know how to make pot roast. She has a theory of cooking, and she uses this theory to understand new recipes, to predict how and when it is reasonable to substitute ingredients, and to understand why something went wrong with a recipe. In sum, she uses it to learn.

Now cooking may seem like a frivolous matter for cognitive science to concentrate on (that, of course, depends on how you feel about food). But if we look at other competencies that people have – communicating, planning, making decisions, designing, and so on – it has to be admitted that these form a large part of everyday intelligent behaviour. In our view, the evidence is overwhelming that these fundamental intellectual abilities are based on coherent bodies of knowledge in the same way that cooking is.

KNOWLEDGE AND INTELLIGENCE

The bottom line is that intelligence is a function of knowledge. One may have the potentiality for intelligence, but without knowledge, nothing will become of that intelligence. Acquiring knowledge is thus at the heart of intelligence. In AI, acquiring knowledge is the main issue. We attempt to build machines that can acquire knowledge on their own, but more often must hand-code that knowledge. There is no intelligent behaviour that is not knowledge-dependent.

It therefore follows that education means providing knowledge to children. The question is, knowledge of what sort? The difference between animals and humans is in the kinds of knowledge that they can acquire. We recognise, for example, that humans can acquire knowledge about mathematics, philosophy, or psychology, and animals cannot. What we fail to recognise is that just because humans can acquire such knowledge, it does not follow that such knowledge forms the crux of what they should acquire.

It makes a lot more sense, once one has appreciated the AI position, to concentrate on teaching children the kinds of knowledge that they must acquire in order to be intelligent. To put this another way we do not, as AI researchers, need to teach machines about philosophy or mathematics. We do, on the other hand, need to teach them scripts, how to generalise, how to abandon a script, how to understand what is the same and what is different, how to characterise an experience, how to deal with an exception. These are the things that are really necessary for intelligence, and yet these are the things that are consistently left out by the school system. Why are they left out? Two important reasons come to mind. First, educators did not know enough about human psychology and the lessons learned from AI to put them in. Second, the philosophies of the Chomskys and the Searles of the world have implicitly prevented these things from being seen as worthwhile, much less critical, subjects.

The major lessons to be learned from AI involve the kinds of knowledge upon which intelligence depends. To make children more intelligent, which we take as the ultimate purpose of education, we must take seriously the lessons of AI. At the same time we must understand that the question 'What makes us different from them?' is of great significance in the debate about education.

This question has two manifestations. For Chomsky and others the question means 'How are humans different from animals?' It should be clear that it is not the form of the knowledge that differs between animals and humans, but the content of that knowledge. Humans differ from animals in what they know and what they need to know, and this determines what they can learn and what they need to learn. This difference ought to tell us a great deal about what we need to teach to children but, unfortunately, it has not.

The philosopher's version of this question is 'How do people differ from machines?' In fact, this would be an interesting question if machines were more advanced than they are. The question of machine consciousness is rather premature, considering how little machines know. If consciousness involves at all to know that you know, then since machines know so little, it is much too early to even consider the issue.

Unfortunately, when we talk about artificial intelligence, we often do not talk in the same way that we would if the subject were human intelligence. Humans exhibit a range of behaviours that cause those in a position to judge to feel competent to assess their intelligence. However, when the subject is computers and their intelligence, the standards seem to change. No one doubts the basic intelligence of a human who is playing chess, solving a complex problem, or summarising a *New York Times* article, but give that ability to a computer and critics everywhere claim that the machine is just following rules that it does not understand. These arguments have led us to claim that these critics are 'fleshists'. Put behaviour in flesh, and they presume intelligence; put it in plastic, and that same behaviour could not possibly be intelligent.

Perhaps surprisingly, in spite of this argument, we believe that these critics are right. Machines that exhibit intelligent behaviours are not necessarily intelligent. The critics are right, but for the wrong reasons. The argument is not that intelligent machines don't feel intelligent, nor does it rely on eloquent affirmations of the mysteries of the human condition. The argument is not whether there are elegant underlying principles that intelligent machines have failed to embody, nor does it entail grand statements about what machines could never do based upon misconceptions about the capabilities of such machines. And, obviously, the argument is not about competence versus performance.

Most discussions fail to centre on the only real issue, namely, what exactly is intelligent behaviour? Or, to put this another way, what is it that humans do that exhibits their intelligence? The answer is not one that can be stated in terms of learned behaviour that may be spoon-fed to a machine. The answer has to be formulated in terms of learning mechanisms that would allow a machine to learn intelligent behaviours by itself in the way a human would, or in terms of basic behaviours that humans come with naturally, that they do not have to learn.

To put this more simply, it is irrelevant for the arguments about the possibilities for machine intelligence to discuss how well a machine plays chess. What matters for this argument is how the machine learned to play chess. For AI researchers it does matter that a machine plays chess, for by building such machines we can learn quite a bit about the mechanisms involved in playing chess, and this is an important part of AI. However, the ultimate behaviour is irrelevant for the possibility argument unless that behaviour were learned the way humans learn chess, which is not something that has yet happened in AI. On the other hand, there could be behaviours that AI people built into a machine directly that would be germane to this argument if those same behaviours were innate in humans. Otherwise, AI must contribute to explaining how humans learn, and thus how a machine might learn in similar circumstances, in order to make any claims about machine intelligence. The point is, any such argument must accept in principle that humans are defining the game and are thus defining what intelligence looks like. For humans, the idea of intelligence means possessing the ability to learn the way that people do.

Thus, AI people have three tasks to accomplish before anyone ought to take seriously the claim that they have accomplished anything in this regard. They need to identify fundamental behaviours that constitute intelligence. They need to differentiate learned from innate behaviours. And they need to build programs that can learn the learned behaviours starting from the built-in behaviours. This ought to be the critic's argument. Now let's talk about why it is not.

There are two main reasons that this argument has not been made by the assembly of self-appointed AI critics. First, one has to have understood enough about human intelligence and enough about AI to have made such a criticism. The critics haven't made these arguments because they haven't been wrestling with the issues of what intelligent behaviours are and how to reproduce them. Second, they have not made the arguments because to do so would allow that artificial intelligence was possible in principle – and this they cannot do because the existence of such intelligent machines would deny some of their basic tenets. The worst part of this, and the point of this chapter, is that those basic tenets are at odds with the idea of enhancing human intelligence as well. These critics are not only anti-AI; whether they acknowledge it or not, they are also anti-education.

FURTHER READING

Hammond, K., *Case-based Planning: Viewing Planning as a Memory Task.* San Diego: Academic Press 1989.

Lakatos, I., *Proofs and Refutations.* Cambridge: Cambridge University Press 1972.

Riesbeck, C. and Schank, R., *Inside Case-based Reasoning.* Hillsdale, NJ: Lawrence Erlbaum 1989.

Schank, R., *Dynamic Memory: A Theory of Reminding and Learning in Computers and People.* Cambridge: Cambridge University Press 1982.

Schanck, R., *Tell Me a Story: A New Look at Real and Artificial Memory.* New York: Scribner 1990.

Schank, R. and Abelson, R., *Scripts, Plans, Goals, and Understanding.* Hillsdale, NJ: Lawrence Erlbaum 1977.

Sussman, G. *A Computer Model of Skill Acquisition.* New York: American Elsevier 1975.

Waterman, D. and Hayes-Roth, F. (eds.), *Pattern-directed Inference Systems.* New York: Academic Press 1978.

Mathematical intelligence

ROGER PENROSE

What is mathematical intelligence? Is there anything that is essentially different about the way that we reason mathematically, from the way in which we think generally? Is mathematical intelligence different from any other kind of intelligence?

I feel certain that there is no fundamental difference between mathematical and other kinds of thinking. It is true that many people find it difficult to cope with the abstract type of thinking that is needed for mathematics, whilst finding comparatively little difficulty with the equally convoluted judgements that are involved in day-to-day relationships with other human beings. Some kinds of thinking come easily to certain people, whereas other kinds come more easily to others. But I do not think that there is any essential difference – or that there is more difference between mathematical thinking and, say, planning a holiday, than there is between the latter activity and understanding a music-hall joke. Human mathematical intelligence is just one particular form of human intelligence and understanding. It is more extreme than most of these other forms in the abstract, impersonal, and universal nature of the concepts that are involved, and in the rigour of its criteria for establishing truth. But mathematical thinking is in no way removed from other qualities that are important ingredients in our general ability for intelligent comprehension, such as intuition, common-sense judgement, and the appreciation of beauty.

What, after all, is intelligence? What is thinking? There is a prevalent viewpoint in current philosophising that holds that whatever it is that in detail constitutes the physical activity underlying our thought processes, it cannot, in effect, be other than the carrying out of some vastly complicated calculation. The relevant actions of our brains, so it is argued, are simply to provide our bodies with a very effective control system – a control system that could in principle be effected by a computer, if only one knew enough of the details of those computational procedures that the brain actually carries out. One might well imagine that, in accordance with this view, such an underlying computational basis to our thinking ought to be most manifest with mathematical thinking. For is not mathematics a computational activity *par excellence*? Indeed, it is not! It is one of my purposes here to emphasise that there is a great deal of what is essential in mathematical thinking that is not of a computational character. Indeed, it turns out that it is possible actually to demonstrate that there is something in our mathematical understanding – in our insights as to mathematical truth – that eludes any computational description whatever. It is the very precision and the universal character of mathematical argument that allows such a demonstration to be possible. But the conclusion is in no way restricted to an intelligence that relates merely to mathematical thinking. As I have argued, there is nothing essential that separates mathematical from other types of thinking, so our demonstration that mathematical understanding is something that cannot be simulated in a computational way can be thought of, also, as a demonstration that understanding itself – one of the most essential ingredients of genuine intelligence – is something that lies beyond any kind of purely computational activity.

MATHEMATICAL VISUALISATION

What are we doing when we conjure up in our minds the image of some mathematical structure? Are we performing some internal calculation, like those that lead to the impressive computer graphic displays with which we are now so familiar? Perhaps our brains are

Figure 1 A regular dodecahedron

acting out something like the computational procedures that give rise to what is called 'virtual reality', whereby an entire three-dimensional seemingly consistent structure, such as a non-existent building, can be visually presented to a human subject, through the agency of a pair of special stereoscopic goggles. The detailed scene that each eye perceives is the result of a complicated calculation performed in 'real time' so that the structure appears to remain consistent no matter how the subject turns his head or moves his body. Are we doing something similar when, in our 'mind's eyes' we conjure up some consistent mental image of a three-dimensional object, whether real or entirely imagined?

I shall argue that we are actually doing something very different from this. Let us consider an example. Figure 1 is a photograph of a regular dodecahedron. With some effort, it may be possible for us to rotate this image to a different orientation. In fact, we may feel that we have some conception of the object as an actual three-dimensional structure rather than as something that needs a particular vantage

Figure 2 A cube

point from which to view it. Many people would find considerable difficulty in visualising an entire dodecahedron, but the cube depicted in the photograph of Figure 2 is a good deal easier. It is not that hard to transfer the flat image on the page to a 'solid' three-dimensional imagined structure. This may seem to be similar to what is involved with computer graphic displays. In Figure 3, I have provided a sequence of computer images of a regular dodecahedron viewed from successively slightly different vantage points, so that the

Figure 3 Computer pictures of a dodecahedron, from a gradually moving vantage point

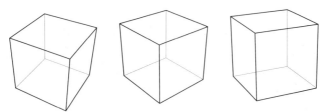

Figure 4 Computer pictures of a cube, from a gradually moving vantage point

dodecahedron appears to rotate just as it would if we slowly move around it. In Figure 4, I have done the same thing with a cube. In each case, there is, inside the computer, a stored representation of the dodecahedron, or cube, that does not change, but the chosen vantage point is gradually altered. Might not our own visual images be something like such stored computer representations?

I think it is unlikely – and to support this contention, let me indicate some fairly clear-cut distinctions. In the first place, the computer displays are far more accurate than anything that can be at all easily achieved by human imagination. Of course, it might well be argued that we are simply being very inefficient and inaccurate in our visualisations, as compared with a modern computer. Indeed, it would not be hard to introduce inaccuracy into our computer simulations, so that they fall to the level of accuracy that would be relevant to any particular human individual. If it were just our inaccuracy that distinguishes our own acts of visualisation from the outputs of computers then my argument would certainly be a very weak one. But visualisation carries with it strong elements of understanding, and it is actual understanding that the computer simulations lack.

To illustrate this point, consider Figure 5. Here I have added some lines to the photograph of Figure 1 to show that a cube can be found inside the dodecahedron, its eight vertices coinciding exactly with a selection of eight from the twenty vertices of the dodecahedron. It is not hard to see that this selection of eight vertices indeed gives us an exact cube. Symmetry considerations alone will tell us this; each face must clearly be an exact rectangle, at least, and the rectangle's sides must indeed be equal since each is a 'diagonal' of one of the equal

Figure 5 Dodecahedron with additional lines to show that a cube sits within it, sharing eight of its vertices

regular-pentagonal faces of the dodecahedron. Can the computer simulation 'see' this fact? This position would be hard to maintain, I feel, but we can at least test it by asking whether or not two of the edges of the computer's proposed 'cube' are equal. (This, in itself, is far from sufficient, of course, but at least it is a start.) When I tried this on two edges, the computer came out with two ten-digit decimal expressions that differed in the final place. When I asked it whether these two numbers were equal, it asserted that indeed they were. (Apparently, the program allows for some round-off error.)

This is hardly a convincing demonstration that the computer in any sense 'knew' that there is an absolutely exact cube in the regular dodecahedron. Rather, it seems to me, it establishes the contrary conclusion, that all this particular computer 'visualisation' can do is come up with approximations, albeit approximations with nine or ten figures of accuracy. It has no way of reaching the exact conclusion that our own visualisations – and accompanying understandings – are capable of: that indeed our proposed cube is geometrically precise.

In fact, being wise after the event, a computer programmer might replace the particular way in which the dodecahedron is stored in the machine by another one whereby exact information about distances and angles could be retrieved on demand. It would then be possible for the machine to give a correctly affirmative answer to our question about whether the suggested cube is indeed an exact one. Actually, being wise after the event, there would be an even easier way: the computer could be instructed simply to answer 'yes' to this particular question about the cube! The trouble with this, of course, is that the computer itself could in no way be said to possess any mathematical understanding of the exactness of the cube. It would simply be parroting the information that its programmer had provided it with, and no-one would argue that any understanding of the exactness lay with the computer rather than with its human programmer.

One might try to do better than this, of course, and perhaps equip the computer with the complete system of axioms for three-dimensional Euclidean geometry. It could then try to ascertain whether a given statement, such as the exactness of the cube referred to above, could be deduced from these axioms. In this way it could, in principle at least, provide the correct answers to many geometrical problems. Of course, it might still be questioned whether what the computer does bears any relation to what a human mathematician does when understanding that a geometrical statement is actually true. That human understanding has to do with a belief in the validity of those intuitions – based to a good extent on symmetry considerations – underlying the very choice of the axioms themselves. The issue is a somewhat delicate one for there are valid geometric axiom systems that are distinct from those of Euclid. Indeed, when I later present powerful arguments in support of the thesis that our insights and understandings are not things that can be reduced to computation, it will be necessary to turn away from geometry, and to address the issue of computation directly, where it will be our intuitions concerning the natural numbers $0, 1, 2, 3, 4, 5, \ldots$, rather than geometrical forms, that will be the subject of our deliberations.

But before turning to such matters, let me give credit to what might be called one of the early 'success stories' of Artificial Intelligence. In the early 1960s, H. L. Gelernter programmed a computer to derive propositions in Euclidean plane geometry from the axioms with which it had been initially provided. When the computer came up with its proof that if a triangle has two equal sides, then the angles opposite to those sides are also equal, Gelernter was startled. For the computer's proof had been unknown to him, and it was considerably simpler than that given by Euclid. The computer's argument was this (see Figure 6): since AB=AC, the triangles ABC and ACB are congruent (side-side-side); therefore <ABC equals <ACB, QED! In fact this argument was not new. (It was given in the fourth century AD by Pappus.) But it was undoubtedly a striking fact that a computer could come up with something so elegant and unexpected.

In this example, the computer's success arose because its 'blind' rule-following stopped it from being distracted from the seeming absurdity of its own argument. No doubt there are many other situations in which human mathematicians have been similarly distracted from seeing arguments that they should have seen. However, this particular example is one for which the chain of reasoning is very short, and it is not hard to find it by means of a mindless search. When the derivations from axioms are long and complicated, as they tend to be with mathematical arguments of considerable sophistication, then the 'mindless search' method becomes hopelessly inefficient.

Figure 6 An isosceles triangle; the angles at B and C are equal – as proved by Gelernter's computer program, since ABC and ACB are congruent

Insight and understanding are necessary ingredients, if the search is to become manageable.

PROOF BY GEOMETRICAL INSIGHT 1: FAREY FRACTIONS

Choose a reasonable-sized natural number n (say $n=9$) and write down, in order of size, all the fractions from 0 to 1, expressed in their lowest terms, whose denominators do not exceed n (here with $n=9$)

0/1, 1/9, 1/8, 1/7, 1/6, 1/5, 2/9, 1/4, 2/7, 1/3, 3/8, 2/5, 3/7, 4/9, 1/2, 5/9, 4/7, 3/5, 5/8, 2/3, 5/7, 3/4, 7/9, 4/5, 5/6, 6/7, 7/8, 8/9, 1/1

Such an array is referred to as a sequence of *Farey fractions*. Let me point out one remarkable property that such a sequence possesses. If we take the difference between any pair of consecutive fractions in the list, then we find that the numerator is always 1. For example

$$2/5 - 3/8 = (16-15)/(5 \times 8) = 1/40$$
$$4/7 - 5/9 = (36-35)/(7 \times 9) = 1/63$$

How do we know that this must always be the case? We must find a way of convincing ourselves that if a/b and c/d are consecutive fractions in the list, then

$$ad - bc = 1$$

(since $a/b - c/d = (ad-bc)/bd$). To do this, we can use a geometrical argument. Let us imagine each fraction a/b to be plotted as the point (a,b) on a graph, where we also take the points representing successive fractions to be joined by straight-line segments. Figure 7 gives an accurate representation of all this, for $n=9$, but the figure is rather crowded and it will be much clearer if we use the inaccurate representation of part of the sequence that is illustrated in Figure 8. (It is a striking fact about mathematical diagrams, and the attendant visualisation of abstract concepts – which is indeed what is going on here – that an inaccurate image, if it is inaccurate in an appropriate way, may often give more of an insight as to what is going on than

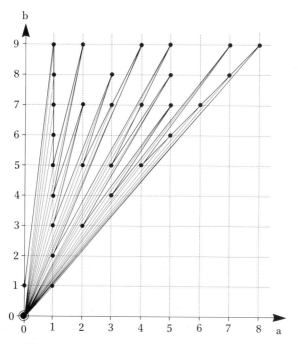

Figure 7 Farey fractions illustrated geometrically

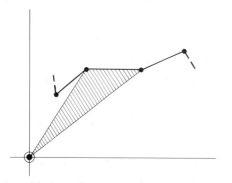

Figure 8 The area of the shaded triangle is $\frac{1}{2}$ unit

an accurate one!) We wish to show that $ad-bc=1$. It is a well-known formula of coordinate geometry that the *area* $\triangle$ of the triangle whose vertices are $(0,0)$, (a,b) and (c,d) is given by

$$\triangle = (ad - bc)/2$$

so what we must establish is that this area is precisely 1/2 (Figure 9).

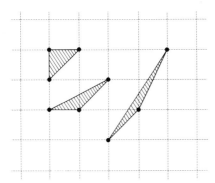

Figure 9 Triangles whose vertices are on lattice points and which contain no other lattice points, either internally or on an edge, are each of area = $\frac{1}{2}$ unit

In fact, it is a theorem (essentially a special case of one due to Minkowski) that if a triangle's vertices all lie at integer lattice points (the points (x,y) where x and y are integers) and if it has the property that it contains no other lattice point either in its interior or on its edges, then it must have an area equal to 1/2. Thus, what we need to show is that our triangle indeed has this property. This is actually not hard to see from the fact that the two fractions a/b and c/d are both in their lowest terms (so that there are no further lattice points on the sides of the triangle out from the origin $(0,0)$), and from the fact that all fractions with denominators no greater than n have been included – a fact that would be contradicted if our triangle constructed from two successive fractions a/b and c/d contained any other lattice point.

It remains to establish the aforementioned theorem. One way of doing this (Figure 10) is to consider a sequence of transformations whereby the triangle is successively moved without changing its area until it becomes just half of a lattice square. Each transformation consists of taking one of the vertices and moving it parallel to the opposite side, until it reaches another lattice point that lies closer to the other two vertices than before. I shall not bother with the full details of this here, but I hope that the rough idea is clear.

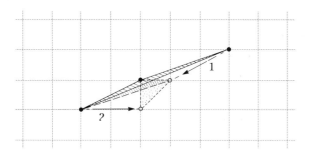

Figure 10 The fact that the area is ½ unit can be proved by successively displacing a vertex to another lattice point closer to the opposite side, in a direction parallel to that side, until a diagonally bisected square is obtained

PROOF BY GEOMETRICAL INSIGHT 2: HEXAGONAL NUMBERS

My next example involves what are called hexagonal numbers

1, 7, 19, 37, 61, 91, 127, . . .

namely the numbers that can be arranged as regular hexagonal arrays (excluding the vacuous array):

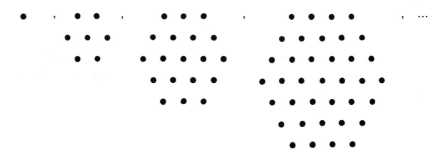

These numbers are obtained, starting from 1, by adding successive multiples of 6

6, 12, 18, 24, 30, 36, . . .

as we see from the fact that each hexagonal number can be obtained from the one before it by adding a hexagonal ring around its border:

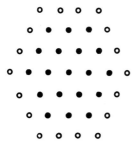

For the number of spots in this ring is a multiple of 6, the multiplier increasing by 1 each time, as the hexagon gets larger.

Now let us add together the hexagonal numbers successively, up to a certain point, starting with 1. What do we find?

$$1=1,\ 1+7=8,\ 1+7+19=27,\ 1+7+19+37=64,\ 1+7+19+37+61=125$$

The numbers 1, 8, 27, 64, 125 are all *cubes*. A cube is a number multiplied by itself three times

$$1=1^3=1\times1\times1,\ 8=2^3=2\times2\times2,\ 27=3^3=3\times3\times3,\ 64=4^3=4\times4\times4,$$
$$125=5^3=5\times5\times5,\ \ldots$$

Is this a general property of hexagonal numbers? Let's try the next case. We indeed find

$$1+7+19+37+61+91 = 216 = 6\times6\times6 = 6^3$$

I am going to try to convince you that this is always true. First of all, a cube is called a cube because it is a number that can be represented as a cubic array of spots

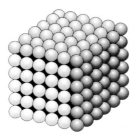

Figure 11 A cubical array of spheres

I want you to try to think of such an array as built up successively, starting at one corner and then adding a succession of three-faced arrangements each consisting of back wall, side wall and ceiling, as depicted thus:

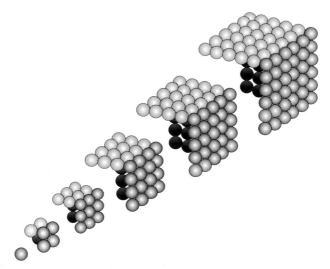

Figure 12 The cubic array is separated into a succession of layers consisting of side wall, back wall and ceiling, each of which is viewed from a long distance off

Now view this three-faced arrangement from a long way out, along the direction of the corner common to all three faces. What do we see? A hexagon:

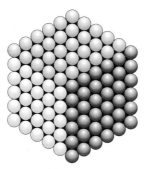

Figure 13 Each layers appear as a hexagonal arrangement – a hexagonal number

The spots that constitute these hexagons, successively increasing in size, when taken together, correspond to the spots that constitute the entire cube. This establishes the fact that adding together successive hexagonal numbers, starting with 1, will always give a perfect cube.

PROOF BY GEOMETRICAL INSIGHT 3: RULES OF ARITHMETIC

This last example again shows the power of mathematical (geometrical) visualisation. Let us try something else; this time a good deal more elementary. How do we know that, for two natural numbers a and b, we always have

$$a + b = b + a ?$$

For this, all we need to do is visualise a collection of things (imagined to be 'a' in number) to which we add another collection of things (imagined to be 'b' in number). The total number of things altogether is clearly the same whichever order we add them in, so the required result follows.

This example is very trivial, and it is perhaps not quite 'geometrical' in the ordinary sense, but I think that it represents a genuine mathematical insight from an act of imagination. One might try to argue that this insight is really just an aspect of our 'experience' of the persistence of objects in the world. For example, apples and oranges do not just disappear when placed in a box, nor do they magically appear within the box. But there is more to it than this. None of us has ever directly experienced precisely 88990012345 objects or 60606999931 objects, yet we would have no doubt that indeed

$$88990012345 + 60606999931 = 60606999931 + 88990012345.$$

Our experiences with apples and oranges merely act as a guide towards our mathematical insights. Those insights are genuine abstractions that are valid methods of reasoning about abstract mathematical objects. (As always, in mathematics, one must be

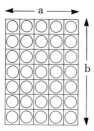

Figure 14 A rectangular array of *a* columns of *b* objects gives the same total number as would *b* columns of *a* objects – by rotation of the figure

extremely careful, however. There are kinds of *infinite* numbers for which the rule $a+b=b+a$ fails – but that's another story!)

Let's try another rule of elementary arithmetic:

$$a \times b = b \times a$$

This tells us that if we are to imagine taking '*a*' collections of objects, where each collection contains '*b*' objects, then the total number would be unaltered if we did the same thing but with '*a*' and '*b*' interchanged. Stated this way, it is not really obvious; but if we imagine our collections to be arranged as a succession of '*a*' columns, where each column contains '*b*' objects, then the symmetry becomes obvious (Figure 14). We can imagine rotating the resulting rectangular array through a right angle to achieve this. Even easier than this (if we have mental difficulty in visualising the rotation) is simply to read our array off the other way around rather than mentally rotating it, that is as row-by-row rather than column-by-column. This works just as well for, say,

$$97666000011 \times 777708999 = 777708999 \times 97666000011$$

as for $5 \times 7 = 7 \times 5$, even though we cannot precisely visualise collections of things that represent these actual individual numbers.

How about the associative law

$$(a \times b) \times c = a \times (b \times c)$$

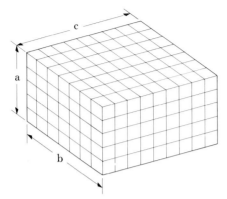

Figure 15 A three-dimensional rectangular array shows that
$(a \times b) \times c = a \times (b \times c)$

We can see this one by imagining a three-dimensional rectangular array which we read off in two different ways: see Figure 15. Again this works just as well for, say

$$(97666000011 \times 777708999) \times 83383302222$$
$$= 97666000011 \times (777708999 \times 83383302222)$$

as for $(5 \times 7) \times 9 = 5 \times (7 \times 9)$, even though these large numbers cannot be individually visualised precisely.

An important thing to note about these 'visualisations' is that they are not really pictures of actual things in space, but of something much more abstract. It really does not matter whether 'actual' space has the necessary accurately Euclidean structure, or that it extends outwards far enough so the particular numbers that we wish to represent can be realised in terms of actual objects within that space. (For example, if we wished to represent the number $10^{10^{1000}}$, which is a perfectly good number, still subject to the same algebraic laws as the numbers that we can directly visualise, then we could not do so within the 'observable universe' with actual physical objects.) Nevertheless, our simple visualisations are indeed sufficient to provide the necessary insights that can convince us that the algebraic properties that we have been considering are actually true of all natural numbers, no matter how large they may be.

Now, let us consider the relation

$$((a\times b)\times c)\times d = a\times(b\times(c\times d))$$

Can we be sure that this one is universally true? Now we use a different method. It is not much good to use a four-dimensional array, which is what would be needed if we were to try to visualise the relation directly. We have no direct experience of four spatial dimensions, so our geometrical intuitions are not of much immediate use in this case. Instead, we can turn to algebra and deduce this last relation from what we have established previously. First

$$((a\times b)\times c)\times d = (a\times b)\times(c\times d)$$

by the earlier relation (but with $a \times b$, c, d, in place of a, b, c, respectively); then

$$(a\times b)\times(c\times d) = a\times(b\times(c\times d))$$

again by this earlier relation (but now just with $c\times d$ in place of c). The result follows by combining these two.

Algebra provides a very useful means of replacing our direct insights by calculational procedures. We do not now have to think of what our expressions actually 'mean'. We can now just calculate! In fact this is somewhat overstating the case I want to make. The effective use of algebra often requires a good deal of understanding, subtlety, and even artistry. But in many ways, it is the power of a good calculus (like ordinary algebra or, indeed, ordinary arithmetical notation) that, to a considerable degree, it enables understanding to be temporarily suspended and replaced by blind calculation. It is in this facility for blind calculation, rather than understanding, that computers can far exceed the capabilities of even the most effective of human experts.

CAN UNDERSTANDING ALSO BE
REDUCED TO BLIND CALCULATION?

What about the human faculty of understanding? Can we be sure that it is not itself some kind of calculational activity? Presumably anyone who believes that genuine Artificial Intelligence is possible must also believe that the quality of understanding can be artificially simulated – for understanding is surely an essential part of genuine intelligence. The present meaning of the term Artificial Intelligence (or 'AI') is that it is something according to which the artificial simulation is indeed performed calculationally, by which I mean by the use of electronic computers. Although it is possible to imagine that, in the future, some means of 'artificial simulation' which is different from anything that can be achieved calculationally might be introduced, I shall stick to the standard 'calculational' meaning of the term here. Accordingly, those who believe that genuine AI is possible must believe that the quality of understanding can indeed be simulated calculationally.

Before giving my main reason for disbelieving this possibility, I should be a little clearer about what I mean by the term 'calculational'. In effect, as I indicated earlier, I mean anything that can be performed by a modern general-purpose computer. This is perhaps not very precise as a mathematical definition. What I really mean, technically, is anything that can be performed by a Turing machine. However, I appreciate that many readers will not know precisely what the term 'Turing machine' means, so it is easier just to refer to a modern general-purpose computer, where we require this 'computer' to be a mathematically idealised concept. The idealisations that we require are that the computer never makes mistakes, that it can continue indefinitely without ever wearing out, and that it has an unlimited storage capacity. If you think of this last idealisation as being a little unreasonable, just imagine that it is always possible to add more storage capacity (that is 'memory') to the computer whenever it runs out.

An important point to make here is that actions of things like 'parallel computers' and (artificial) 'neural networks' (or 'connection machines'), that we hear quite a lot about these days, are all included in what I mean by 'calculations'. For some types of problem, a computer designed according to what is called a 'parallel architecture' may be much more efficient or much faster than an ordinary serial computer, but there is no difference between the two in principle. Likewise, systems like neural networks – referred to as 'bottom-up' systems – which improve upon their performance by a 'learning' process aimed at optimising the quality of their output, are also calculational. A bottom-up action contrasts with the standard 'top-down' calculational procedures in that the latter operate according to an algorithm that is known to work correctly for the class of problem that it is concerned with, whereas with bottom-up action no such algorithm is given beforehand and, instead, a means is initially provided whereby the system is to improve its performance as it gains experience. This is still a calculational procedure (and therefore still an 'algorithm'), however, because the very means whereby the system is to improve its performance is itself given by a set of calculational rules. From the point of view of the present discussion, the essential difference between a bottom-up and a top-down system is that the former is only an approximate means of obtaining the required answers even though it may sometimes be a very effective one.

A simple test for deciding whether a system is calculational is to ask: can it be run on an ordinary computer? If it can be then the system is indeed a calculational one. In fact, as far as I am aware, most of the (artificial) neural network systems that have been constructed to date are actually, as they stand, simulations run on ordinary computers – so that there is no question but that they must be calculational! (Perhaps it is intended eventually to construct some special electronic hardware, on which the neural network system would be run much more efficiently, but that makes no difference to the fact that such systems are always calculational in nature.)

I now wish to present an argument that effectively demonstrates that mathematical understanding is not a calculational activity. The argument is based on a form of the famous theorem of Kurt Gödel, that he proved in 1930, but where I also call upon some later ideas introduced mainly by Alan Turing in about 1935. I shall not need any of the technical details of Gödel's argument, and although the argument may be found to be somewhat confusing, it does not use any difficult mathematics.

The type of calculation that we shall be concerned with will be an operation that can be performed on a natural number. The action of some calculation C on a natural number n is written $C(n)$. We may think of C as being given by a computer program, where after feeding the program into our computer, we then supply the computer with the number n, which the computer operates on to produce the answer. (Technically, C may be thought of as a Turing machine, and $C(n)$ is the action of that Turing machine on the natural number n.) I shall not be very concerned here with the actual result of the calculation, but mainly whether or not it ever eventually stops.

Let us consider some examples. One possible calculation would be to form the square n^2 of the natural number n. This particular calculation encounters no problem about eventually stopping, since the square of any given natural number can certainly be formed in a finite time. (Recall that there is to be no limit on the computer's storage space.) More subtle is the following example of a calculation, which depends on the given natural number n.

> *Find the smallest natural number that is not the sum of n squares.*

Our calculation would proceed, trying the natural numbers in turn: 0, 1, 2, 3, 4, 5, . . . ; until it finds one that is not the sum of n squares. To get the idea, let us first see how this works for $n=2$. We start with 0, and find that 0 is indeed the sum of two squares, namely $0=0^2+0^2$. We must move on to 1, and find that although 0^2+0^2 doesn't work, we do find that indeed $1=0^2+1^2$. Thus, we must move on to 2, finding that although 0^2+0^2, or 0^2+1^2, or 0^2+2^2 do not work, we indeed have $2=1^2+1^2$. Moving on to 3, we find that none of 0^2+0^2, 0^2+1^2, 0^2+2^2, 0^2+3^2, 1^2+1^2, 1^2+2^2, 1^2+3^2, 2^2+2^2, 2^2+3^2, or 3^2+3^2, will work (cutting the calculation off when the number to be squared reaches the number to be summed to – though we could be more efficient, cutting things off earlier). Thus we find 3 as the smallest number that is not the sum of two squares. We could now try this all over again with $n=3$, finding, in this case, that the number 7 is the smallest that is not the sum of three squares. We can also go back and test the case $n=1$, finding that 2 is the smallest number not the sum of one square (and examining the logic involved in the case $n=0$, we find that 1 is the smallest number not the sum of zero squares).

Now let us consider $n=4$. Our calculation proceeds, finding

$$0=0^2+0^2+0^2+0^2,\ 1=0^2+0^2+0^2+1^2,\ 2=0^2+0^2+1^2+1^2,\ \ldots$$
$$6=0^2+1^2+1^2+2^2,\ 7=1^2+1^2+1^2+2^2,\ 8=0^2+0^2+2^2+2^2,\ \ldots$$
$$23=1^2+2^2+3^2+3^2,\ 24=0^2+2^2+2^2+4^2,\ \ldots$$
$$71=2^2+3^2+3^2+7^2,$$

and so on.

It seems never to stop at all! In fact it never does stop. According to a famous theorem first proved in the eighteenth century by the great French mathematician Joseph L. Lagrange, *every* number is, indeed, the sum of four squares. It is not such an easy theorem. Even Lagrange's contemporary, the great Swiss mathematician Leonhard Euler, a man of astounding mathematical insight, originality and productiveness, had tried but failed to find a proof, so I am certainly not going to trouble the reader with the details of Lagrange's argument here.

Instead, let us try a calculation that is very much easier to see never to stop.

> *Find the smallest odd number that is the sum of* n
> *even numbers.*

The poor computer that is set mindlessly upon this task will certainly never complete its work, no matter what n is – because even numbers always add to even numbers.

I have given some examples of calculations, some of which will eventually terminate to produce an answer and some of which continue for ever. How are we to decide which of these two possibilities will occur in any particular case? When a calculation does not ever stop, by what means can we ascertain this fact? We have seen that this may be hard, as was the case with Lagrange's theorem, but sometimes it is easy, as in the last example. Are mathematicians themselves using some calculational procedure in order to ascertain that non-stopping calculations actually do not stop?

Let us imagine that they do use such a procedure, and, moreover, they are aware of the nature of this procedure and of the fact that the procedure that they use is sound – that is to say, that it does not erroneously come to the conclusion that a calculation does not stop when in fact it does. It will not be necessary to assume that this procedure can, in every case, ascertain that a non-stopping calculation does not in fact stop.

Let us call our putative procedure A. Then when A is presented with a calculation C and with the number n on which C acts, it will be set into action. If the calculation A itself successfully comes to a halt, then it will have decided that C(n) does not in fact terminate. (Note that A is not a procedure for deciding that calculations *do* terminate. We might have some other procedure B for that kind of decision. If we want to incorporate B into A, we can do so by employing the device of putting A into a 'loop' whenever B successfully comes to its conclusion, thus making sure that the 'A' used in the argument will actually not terminate when the calculation does. This is just a technical point. I mention it only because sometimes people are disturbed if A ignores arguments that show that a calculation will stop.)

In order to be a little clearer about how a calculation (A) can act on another calculation (C), let us specify the various calculations C by giving each one a separate number. Thus the different calculations will be

$$C_0, C_1, C_2, C_3, C_4, C_5, \ldots$$

where we can think of this ordering as being provided by the numerical ordering of the computer programs that specify these calculations in turn. Technically, C_r could be the 'r^{th} Turing machine' in some standard system of numbering. We can now think of A as a calculation acting on the two numbers r and n, and conclude

If A(r,n) stops, then C$_r$(n) does not stop

The calculation A is still not quite of the form of the other calculations that we have been considering since it acts on two natural numbers, not one. Let us remedy this by considering only the cases for which $r=n$. (This perhaps seems an odd thing to do, but it is the crucial step in Gödel's and Turing's argument, itself taken from the famous 'diagonal slash' of the highly original nineteenth-century mathematician Georg Cantor.) We obtain

If A(n,n) stops, then C$_n$(n) does not stop

Now, $A(n,n)$ is of the form of the calculations that we have been considering, so it must be one of them, say the kth one, and we have

$$A(n,n) = C_k(n)$$

and therefore, putting this in the displayed statement above

If C$_k$(n) stops, then C$_n$(n) does not stop

Taking the particular case $n=k$, we obtain

If C$_k$(k) stops, then C$_k$(k) does not stop

From this we deduce that $C_k(k)$ certainly will not stop (because if it did, then it doesn't!).

The remarkable thing, here, is that although we have ourselves just seen that $C_k(k)$ does not stop, the calculation A is incapable of ascertaining this fact. For $A(k,k)$ is the same as $C_k(k)$, so if the latter does not stop, the former cannot stop either. Thus, A cannot successfully come to the conclusion that $C_k(k)$ does not stop! Since we have actually just established that $C_k(k)$ does not stop, it follows that the mathematical procedures that we use in order to establish that calculations do not stop are not accessible to the calculational procedure A. We note this would apply whatever A is, provided that we know what A is, and we know A to be sound. The inescapable conclusion seems to be:

*Mathematicians are not using a knowably sound
calculational procedure in order to ascertain
mathematical truth*

We deduce that mathematical understanding – the means whereby mathematicians arrive at their conclusions with respect to mathematical truth – cannot be reduced to blind calculation!

DISCUSSION OF THE IMPLICATIONS OF THE GÖDEL ARGUMENT

I should address some of the possible loopholes and objections that various people have made to arguments of the type that I have just given.

A common reaction to the Gödel argument is simply not to take it seriously, for 'how could an argument of that kind possibly have anything to say about the mind?' But although this may be a natural reaction, it is no answer to the argument. If one believes that the conclusion is wrong, then one must find a flaw in the argument.

A worry that people often have is that I have given the argument in terms of a single A, whereas there might be a whole host of calculational procedures that mathematicians use. However, this is not a real objection. There is no difficulty about combining together many different such procedures (even an infinite number of them) into a single 'A', provided that it is a calculational matter to decide which procedure to use. It is only for convenience that I have phrased things as I have, and there is no loss of generality involved.

One common objection is to point out that the Gödel(–Turing) argument is itself something that one could envisage putting on a computer. There is nothing non-computable about generating all the steps of the argument as I have given them and, if we wish, we could include some version of the Gödel argument into our rules for deciding that calculations will not stop. But if we simply adjoin this new argument to the 'A' that we had before, we are really cheating, because that 'A' was already supposed to represent the totality of the means that are available to mathematicians for ascertaining that calculations do not stop. If we accept the Gödel argument as a new means for ascertaining that certain calculations do not stop, then our previous 'A' did not represent that totality. Instead, we should be using some new 'A', say A*, that includes this version of the Gödel argument. But if A* is supposed to represent that totality then we can apply our argument instead to A*, and again we obtain a contradiction. The point is that we cannot put the entire idea of the Gödel argument into calculational form even though we can incorporate certain instances of it into a calculation.

There is a closely related objection that people sometimes try to make against the version of the Gödel argument that I have given. The claim is sometimes made that the argument applies only to the one particular A that has been singled out, and that it is not a general objection against all As. This is a misconception about how the argument is being used, however. The argument has the form, familiar to mathematicians, of a *reductio ad absurdum*, whereby a hypothesis is put forward (here that there is some knowably sound calculational procedure that we use – and that we are calling it A) from which a contradiction is obtained, thereby showing that the hypothesis was false. The argument indeed rules out all such As, not just a particular one.

The Gödel argument is more often phrased in terms of some axiom system F, and in terms of the provability of mathematical results from F. Gödel's most familiar theorem shows that provided F is (and is believed to be) consistent – so it cannot be used to prove that a statement is true and false at the same time – then there are mathematical propositions that are (and are seen to be) true but which cannot be derived from F. The argument is often made that we cannot actually see that these propositions are true unless we can show that the axioms are consistent. I have not phrased my own argument in this way, but have referred to the soundness of the procedure A. If we trust A not to make mistakes, then we see, by the Gödel argument that A cannot represent the totality of our mathematical insights, whatever A might be. Likewise, if we trust F – and this implies that we believe F to be consistent, for otherwise we could use it to prove a nonsense like $1=2$ – then we see that F does not represent the totality of our mathematical insights. If we do not trust F (or A) then it certainly cannot represent our insights!

Another objection I have seen made is that we cannot be sure that the numbers that we are talking about are actually the natural numbers $0, 1, 2, 3, 4, 5, \ldots$, but they might be some funny kind of 'surreal' numbers, where some of the things that would be true for the natural numbers turn out to be false for the surreal numbers. Although I cannot really take this argument seriously – because the numbers that we are talking about *are* the natural numbers, and not anything else – this objection contains, in a sense, the 'nub' of the mystery. How do we know that it is the natural numbers that we are indeed talking about? We cannot, merely by specifying a finite system of axioms or rules, completely distinguish the natural numbers from all the various kinds of 'surreal' numbers. Yet every child knows what the things 0, $1, 2, 3, 4, 5, \ldots$ mean, despite this fact. Somehow we have a direct intuition which tells us what the 'natural number' concept is, given only very inadequate hints in terms of 'two bananas, five oranges, zero socks' and so on.

Let us then accept the apparently inescapable implication of the Gödel(–Turing) argument: *mathematicians do not simply ascertain mathematical truth by means of knowably sound calculational procedures*. There remain the possibilities that they might use unknowable or unsound calculational procedures – or, as is my own belief, that they simply do not just use calculational procedures when they ascertain truth. With regard to the calculational possibilities, I should point out that mathematicians certainly don't *think* that they are using unknowable or unsound procedures in order to ascertain mathematical truth! They are of the opinion that they are perfectly aware of what they are doing when they use whatever methods they use and, moreover, that these methods are perfectly sound. It is undoubtedly true that mathematicians make mistakes from time to time, but these mistakes are recognisable as such. Another mathematician might point out the mistake, or the very mathematician who made the mistake might notice it later. It is not that there are inbuilt errors that mathematicians are completely incapable of seeing as errors. Consciously, the methods that mathematicians use are neither unknowable nor unsound. If they are indeed using a horrendously complicated unknowable calculational procedure X, or an unsound calculational procedure Y, then these things would have to be completely unconscious.

Is it plausible that they are actually using such an X or Y without knowing it? One point should be emphasised here, and that is the apparently universal nature of the criteria that mathematicians use to establish the truth of their results. Suppose that each mathematician used a different X or Y, personal to that particular mathematician, then they would not be able to convince one another of their arguments. We require a universal X or Y that would have to be built into their brains in a way that would be common to all. How could such an X or Y have arisen? It would have to have been by means of the powerful processes of natural selection that Darwin himself revealed to us. But anyone who has glanced at any respectable modern mathematics research journal will realise how far-removed from the activities of the outside world are the deliberations of mathematicians. If it were the horrendously complicated unknowable X, or the complicatedly erroneous Y, that somehow got implanted in our brains, via our genes, it is very hard to see how this could have been by the process of natural selection, a process geared to promote the survival of our primitive remote ancestors. Much more likely is that there is no such X or Y but, instead, it is a non-calculational quality – the ability to *understand* – that natural selection has favoured. This quality is in no way specific to mathematics, but would have been immensely valuable to our ancestors in many different ways, providing a powerful selective advantage. Only incidentally does it turn out that this same quality is what is needed for mathematics.

WHAT UNDERLIES NON-COMPUTATIONAL BRAIN ACTION?

If we accept that we do something beyond computation when we understand, how can this be reconciled with the view that our brains are just physical objects governed by precise physical laws? One way out might be to adopt a mystical viewpoint according to which the behaviour of the mind could not be accounted for simply in terms of the physical brain. Apparently Gödel himself felt driven to this kind of solution.

For myself, I reject mysticism in favour of a scientific explanation. There are various possibilities to consider. For example, perhaps we use a calculational procedure that is continually improving itself. If so, then, we must ask, how is this improvement coming about. If the improvement is itself governed by some preassigned mechanism, then, as was the case with a neural network, it is still calculational. Might the improvements come about via some continual interaction with the environment? But if this is to give us something beyond calculation, it would imply that there is an essential feature of our environment that cannot even be *simulated* computationally. Of course, it may not be feasible to simulate the particular environment of a specific individual, but to suggest that it is in principle impossible to simulate any appropriate plausible environment is to suggest that there is something essential in the physical action of the world that lies beyond calculation. Once that possibility is accepted, then the possibility that our very brains might act according to some non-calculational action must also be allowed.

What about random ingredients? Would they count as 'non-calculational'? In the sense that a strict Turing machine does not allow for such ingredients, their inclusion would, indeed, take us out of calculational activity. However, in practice, purely random ingredients would add nothing useful to pure calculation. In fact, there are many calculational procedures that call for the inclusion of random ingredients, but these are usually implemented in practice by incorporating what are called 'pseudo-random numbers', these being numbers that are generated by some suitably complicated process that gives them the appearance of randomness even though they are not strictly random. For practical purposes randomness gives us nothing useful that cannot be achieved purely calculationally. (Closely related is the behaviour of what are called 'chaotic systems' which have the appearance of randomness even though they are entirely calculational.)

Finally, there is the possibility that, in appropriate circumstances, the actual behaviour of physical systems might be essentially non-calculational, a conscious brain being one example. My personal belief is that this is indeed the case, but there are several speculative elements that are involved in such a belief. First, one must ask where in physics non-calculational action might be found. I believe that such action must be in an area where present-day physics is in need of radical improvement – what is referred to as 'the measurement problem' in quantum theory. Roughly speaking, such an improved theory would supply a more satisfactory link between the micro-level of atoms and molecules (the 'quantum level') and the macro-level of discernable phenomena (the 'classical level'). I believe that brain action will never be properly understood without such a theory. At least something of this nature will be needed in order to explain the non-calculational aspects of mathematical and other kinds of understanding.

FURTHER READING

Boden, Margaret A., *The Creative Mind: Myths and Mechanisms*, London: Weidenfeld 1990.

Broadbent, D. (ed.), *The Simulation of Human Intelligence*, Oxford: Blackwell 1993.

Chou, Shang-Ching, *Mechanical Geometry Theorem Proving*, Dordrecht: Reidel 1987.

Nagel, E., and Newman, J.R., *Gödel's Proof*, London: Routledge and Kegan Paul 1958.

Penrose, Roger, *The Emperor's New Mind*, Oxford: Oxford University Press 1989. Vintage paperback, 1990.

Intelligence in Traditional Music

SIMHA AROM

Tradition might be seen as the process of transmission, within a community that identifies itself as such, of a specific knowledge and of relatively stable forms of behaviour. These define 'symbolic communities' which, as the French anthropologist Jean Molino puts it, are 'groups of individuals sharing some relatively stable features of language and culture, i.e., relatively stable features of their symbolic organisation systems'.

Traditional music is a symbolic production which, like language for a given community, is transmitted from mouth to ear, from generation to generation, and represents a major constituent of the group's cultural identity. Almost all traditional musics share this character: they are transmitted orally. Memory thus plays an essential role. Even in societies which have systems of notation, such as China, India, Tibet, and others, writing only fulfils a mnemotechnic role, as a memory support. It never assumes a prescriptive function. Since orally transmitted musics are not fixed once and for all in writing, there obviously is large scope for improvisation and variations.

Traditional music can either be art music (the French call it *musique savante*), or popular music. As art music, it may be the subject of abstract speculation, of deductions based on acoustic rules, of a constituted body of codes – sometimes written. In such cases, one can properly speak of a theory, since it is presented in explicit form. Such are the art musics of India, China, ancient Greece, Korea, Indonesia, and the Islamic world (mostly Turkey and Iran).

In popular music, whatever its origin, there is no autonomous theory. Informants cannot give a verbal statement of their own rules of procedure, although they can say in practice what is correct and what is not. Those implicit rules proceed from a cultural consensus among the members of the community where the music is practised. Roman Jakobson has best described this situation in another domain: the oral literature of Serbian rhapsodists.

> A Serbian peasant reciter of epic poetry memorises, performs, and, to a high extent, improvises thousands, sometimes tens of thousands, of lines and their meter is alive in his mind. Unable to abstract its rules, he nonetheless notices and repudiates the slightest infringement of these rules.

As mistakes can only be identified by reference to some abstract scheme, or theoretical organisation, one can posit that there is a theory in popular music, but that it is implicit. Such is the case to this day in Africa, in traditional villages composed of ethnically homogeneous communities.

Intelligence can be studied in three ways: the adaptation of an organism to its environment; the complexity of the system of mental structures required by such an adaptation; the individual know-how, that is the ability of an individual to learn and use those complex structures in an appropriate way, according to the circumstances.

Thus the work of the ethnomusicologist can be seen as a systematic study of those three parameters of adaptation, complexity and know-how, by describing

- the consistency of the relationship between the music and its social and symbolic context
- the intrinsic logic of the musical idiom, considered as a formal system
- the know-how, that is the apprenticeship, the capability for discrimination, classification, and accommodation of the people who perform it

I shall refer here to sub-Saharan Africa, mostly to Central Africa, the music of which I have studied for about thirty years. My aim is to show how those people conceive their own music, and what cognitive procedures and strategies are at work in the particular aspect of cultural behaviour that we call music-making. I will first concentrate on the relationship between music and society, then on the internal organisation of the musical system, and finally on the way each individual of the community learns and uses those musical structures, which are part of the fabric of his or her life and identity.

MUSIC AND SOCIETY

The range of traditional African music coincides with the diversity of ethnic groups and sub-groups, as reflected by the diversity of languages or dialects of these groups. Each community that uses a dialect which distinguishes it from all others, possesses in the same way a specific music.

Like language, traditional African music is collective. It belongs to the whole community which is the warrant of its transmission. It is anonymous and undated. It is functional or, more precisely, circumstantial. In most societies it does not involve professional musicians, although the performers are of course more or less specialised. In most circumstances, there is no dichotomy of performers and audience, and any member of the community is a potential participant in musical practice.

Just as in 'learning one's native language', one acquires the foundations of musical knowledge in an empirical manner. At first, the passive infant is imbued with the music which surrounds him. As his motor abilities develop, the child will progressively take part in diverse musical activities, first by rhythmically clapping his hands, then by singing, and finally by trying his skill at an instrument. Institutional teaching is infrequent: it only takes place during initiation rites of various kinds.

As with most African societies the traditional musics of Central Africa above all fulfil social functions. They are woven into the cycle of individual, family and collective existence to such an extent that they are an inseparable and indispensable part of the social and religious life of the community. All events of any significance, and even many everyday occasions, have some ceremony or display attached to them in which music, being an organic part, has a leading role to play.

Furthermore, music is a means of communication and an indispensable intermediary between human beings and the supernatural forces surrounding them: shades of the ancestors, spirits and jinns, in other words, between human beings and their symbolic representations. This is why Central Africans do not consider music to be a purely aesthetic phenomenon, even though they are quite capable of expressing their tastes and making very precise value judgements about both the music itself and the quality of a performance. But aesthetics remains a secondary question and is never an end in itself. Indeed, music only exists here in order to serve clearly defined purposes other than itself. That is why it is invariably a part of a cult, a collective work session, a dance for pleasure on the night of the full moon or, simpler still, a mother singing to soothe her child.

Though they are connected to social and/or religious circumstances, and therefore essentially functional, traditional African musics constitute autonomous formal systems as well. Within the same cultural context, functionality and musical systematics are in fact closely linked. For every occurrence that needs musical support there is a particular repertoire. Each repertoire has a name in the local language, encompasses a specific number of pieces, and is characterised by predetermined attributions of vocal and instrumental roles, as well as by rhythmic or polyrhythmic patterns that the percussion instruments must perform. Each repertoire thus corresponds to a musical category, distinct from all the others. The totality of an ethnic community's music can therefore be presented as a finite set of mutually exclusive categories, named in the vernacular language.

In general, when referring to a repertoire, one uses a generic term that refers to the circumstances of its performance, but also to a series of notions logically connected with it. Thus the Aka Pygmies possess, to my knowledge, fifteen such categories, represented in Figure 1.

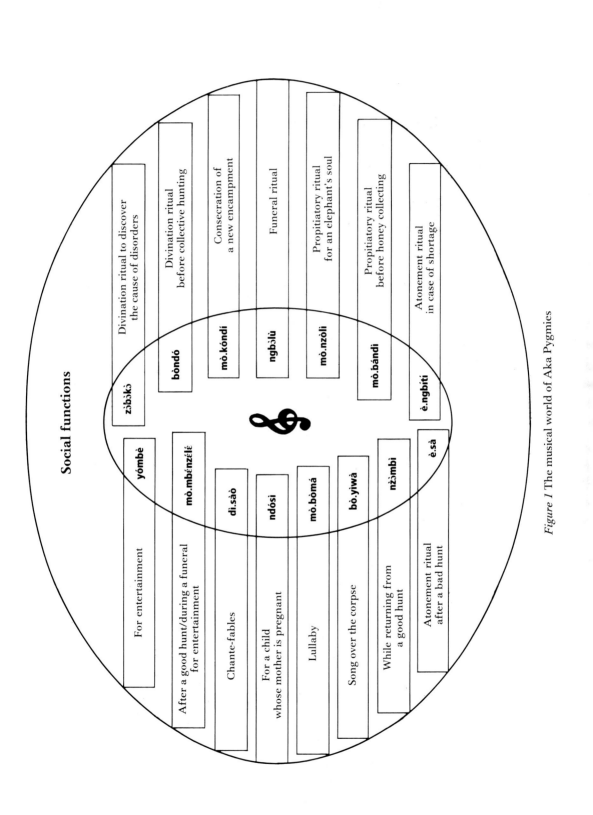

Figure 1 The musical world of Aka Pygmies

(In the central circle, each category is represented by its vernacular denomination, facing the social function or the circumstance it is associated with: thus *mò.kóndí* relates to the inauguration of a new camp.)

Let us look more closely at one of them, namely *bòndó*. For the Aka, *bòndó* is at the same time the name of the divination ritual to discover the cause of disorders, the name of the dance performed by the seer during this ceremony, the name of the seer himself, the name of the ensemble that plays for the dancers, and the name of the specific polyrhythmic pattern to be performed in such circumstances. In addition, *bòndó* refers to the set of songs (each with its own title) that can be performed during this ritual. Finally, if one looks at its literal, non-symbolic meaning, *bòndó* designates a plant, the root of which has hallucinogenic properties, and it is this very root that the diviner, during the ceremony, puts under his tongue, in order to be able to read the immediate future from a big fire.

What we see here exemplifies the Aka Pygmies' conception of a first ordering level of their musical universe. This type of ordering is specific neither to the Aka nor to Central African traditions. In other parts of Africa, both Western and Eastern, each population has an ensemble of musical categories linked with mutually exclusive social and religious functions.

Let us now look at another illustration of the logic that unites symbolic representations and specific aspects of the musical material. In no African language is there a generic term for 'music', but in all of them we find a term for 'song' and another one for 'dance'. These terms suffice to sum up everything we would consider a musical entity. Every piece of music played on a melodic instrument always began by being a song. Therefore the term 'song' is not limited to vocal music, but refers to everything that is considered a piece of music. Moreover, everything that can be designated as a song is of necessity measured in time – that is, underlaid by a series of isochronous pulsations, which may be materialised or not. Thus every sound organisation mapped onto time according to a metrical frame belongs to the musical universe. It is a well-known fact that in most African societies there is no dancing without singing. Hence the concept of dance *ipso facto* implies the concept of song. Conversely, melopeias whose rhythmical unfolding is free, that is non-measured, such as lamentations, are considered to be 'weeping' and not song.

I move now to a case in which the question is precisely whether the presence of non-measured pieces, alternating with measured ones in the same repertoire, is merely a matter of chance, or whether it is culturally significant. The Banda-Dakpa community uses, among other instrumental ensembles, horn orchestras which include a wooden slit-drum. The repertoire of such orchestras consists of some twenty very short pieces. It is performed without interruption, linking strictly measured pieces with others which are not measured.

In order to test the relevance of this contrast, I asked several musicians, separately – and on many occasions – to clap their hands as they listened to a complete recording of their repertoire. A number of pieces were spontaneously scanned by a regular beat, thus actualising the pulsation inherent in each of them. However, as soon as others struck up, the musicians always made the same comment: 'Here you cannot clap.' There was here a clear cultural consensus on the difference between these two types of pieces. Despite many conversations designed to elicit information that could shed light on the meaning of this unusual juxtaposition, no explanation emerged. Finally, the separation of the pieces into measured and non-measured ones revealed that the titles and content of all measured pieces referred to 'real' subjects, like war, hunting, mockery and loneliness, whereas those of the non-measured pieces all referred to beings/animals of avian, aquatic, or terrestrial nature, but always mythical. The difference between the reality of the present and the imaginary and mythical was reflected, in a clearly formal manner, through the temporal organisation of the music. Is this perhaps some reflection of the contrast between the strictness of measured time, the time of the living, and an elastic, fluid time, the time of mythical beings? Whatever the case, the important fact is that symbolic data, even when they initially appeared not to be conceptualised, are indicated here by the presence of a strict distinctive musical feature.

MUSICAL SYSTEMS

The Aka, like many other Pygmy groups, excel in vocal polyphony, which is characterised by the presence of genuine counterpoint. When one listens to the singing of a Pygmy choir, the interweaving of the voices is such that one is tempted to believe that there are as many vocal parts as singers. However, deeper investigation shows that this impression of vocal luxuriance results from the profusion of multiple variations performed simultaneously, all deriving from a limited number of constituent parts. The following questions then arise: how many constituent parts are there? Are they arranged in some hierarchy, and, if so, what is the fundamental part, that is the one on which all the others are based? After collecting, transcribing, and analysing several dozen polyphonic pieces, I was inclined to believe that there were only four constituent parts. With regard to the hierarchy between these, when I returned to the field, I asked the Aka what, for them, was the 'real' song, in other words, the fundamental melody. By way of reply, for each piece, they sang a melodic line which they considered fundamental. These melodies were characterised by the presence of long rhythmical values, and were thus less ornamented than the other parts of the piece. However, I was given no hint of any possible name for this fundamental melodic line. Thus, for my own use, I christened it *cantus firmus*.

On my next visit, during a session intended to verify the role of this *cantus firmus*, a 'productive misunderstanding' (the happy result of a wrong translation), revealed that this part was named in the Aka language *ngúé wà lémbò*, literally 'the mother of the song': that means the melody which bears the song. This led me to suspect that the others might also have specific names. In a few minutes it turned out that there are four constituent parts. Each part has a proper name, which refers to its respective function within the polyphonic edifice. Thus the part which strikes up the song is called *mò.tángòlè* ('that which gives its words'). The part 'which replies to it' (in contrary motion) is called *ó.sêsê* (literally 'below', i.e. subordinate to *mò.tángòlè*). Lastly, the yodelled part, which comes on the top of the three others, is called *dì.yèí*, which means 'yodelling'. It thus immediately became clear why it was impossible to determine (for the analyst as well as for the listener), how many parts were involved. The reason was that in practice each performer may not only sing any part of these four, but may also at any moment switch freely from one to any other. However, according to the Aka conception, of the four constituent parts, two, *ngúé wà lémbò* and *mò.tángòlè* are reserved to men, while *ó.sêsê* and *dì.yèí* are reserved for women. Not only is there a vernacular conceptualisation here for all the parts, but the metaphors used to denote them account perfectly for their status in the polyphonic structure.

Let us now consider how the convergence of numerous criteria, both musical and non-musical, contributes to the establishment of the relevance of a complex conceptualised musical system. For this purpose, I shall refer to another type of horn orchestra, that of the Banda-Linda, who are related to the Banda-Dakpa I have mentioned above. This example shows how external observation is permanently confirmed by cognitive data such as the musical system at work in this repertoire, the way instruments are tuned and used, and the various types of symbolic denominations which will allow the validation of one or another musical parameter, be it on its own or as part of various combinations.

In the Banda-Linda community, horns are only used in a serial set which constitutes an orchestra. These horn orchestras, named *ángó*, and which are intimately related to the rites of passage of young boys, can include from ten to eighteen instruments of different sizes, varying from 30 to 170 centimetres. Their repertoire includes about fifteen pieces; each one bears a title and corresponds to a song specific to that population. Each horn produces a single sound; the interlocking of the individual sounds of the various instruments creates a very complex polyphonic texture, in the technique known as 'hocket'. Each horn plays a precise rhythmic formula on a given pitch; for each piece, each instrument has a specific formula, to which it brings variations during the performance. Going from high to low, the horns are tuned according to the various degrees of the pentatonic scale: G–E–D–C–A. This module is repeated as many times as there are instruments.

The organisation of the pieces of the repertoire takes this order into account: the sequence of entrances, at the beginning of any piece, is always performed in a very strict way, from the highest to the lowest horn. Within a given piece, all the instruments tuned an octave apart depend on the same rhythmic formula (an octave being the interval between any frequency N and 2N). This means, first, that the musical structure of each piece is based on a group of only five instruments, of which all the others are simply duplicates; it means as well that the Linda are perfectly aware of that formal musical organisation. This fact is attested to by the name of each horn: all instruments tuned an octave apart bear the same name. So, whatever their register, those which play the G are called *tété*, those which play the E, *tā*, those which play the D, *hā*, those which play the C, *tútûlé*, and those which play the A, *bɔŋgɔ́*. Each group of five instruments, from the highest-pitched horn downwards, constitutes a 'family', which differentiates itself from the others by its register. As such, it bears a generic name: the five highest horns are called *tūwúlē*, (an onomatopoeic name); the following group, *ngbānjā*, (a kind of rasp), the third one, *āgā*, (the buffalo) and the fourth one, *yāvīrī*, (the storm, the thunder).

		Name of instruments and their notes				
		tété	tā	hā	tútûlé	bɔ̀ngɔ́
Family	Register	G	E	D	C	A
tūwúlē		H 1	H 2	H 3	H 4	H 5
ngbānjā		H 6	H 7	H 8	H 9	H 10
āgā		H 11	H 12	H 13	H 14	H 15
yāvīrī		H 16	H 17	H 18	–	–

Figure 2 Organisation of an eighteen-horn Banda-Linda ensemble. The columns are the degrees of the pentatonic scale; the rows show the octaves defining 'families' of instruments with the same names. H = horn.

We can now understand why the total number of instruments need not be fixed; it is because the musical structure is entirely determined by the performance of the five horns in the higher octave. As soon as the first horn of the second group starts playing, the system of duplicates starts working, without any consequence on the structure of the piece. Thus, five horns are enough to produce the integral musical substance of each piece. The indispensable instruments are always the highest-pitched ones: the 'missing ones' can only be the lowest ones, never the intermediary ones. Since the instruments' entrances follow each other starting from the highest one, each musician takes as reference the formula played by the musician who entered just before him. If an intermediary link were missing, the whole chain would be broken. Thus, the number of horns is unimportant as soon as it exceeds five, and as long as each instrument follows the other without any interruption from the highest one downwards.

Such an organisation incorporates many relevant features which belong to different orders. It shows that the Banda-Linda are aware of the notion of scale, since the instruments corresponding to each degree of the scale have a specific name. They are also aware of the notion of octave, since the instruments tuned an octave apart bear the same name. Finally, the duplication, octaves apart, of the same formulas shows that they are aware as well of the internal organisation of the musical material itself. As the designations of instruments are concomitant with their functions in the musical structure, it is possible to infer from the name of a horn which formula it plays in one or another piece of the repertoire, and reciprocally, it is possible to deduce the name and function of any horn from the formula it plays, since there is an identity of name and formula between all instruments tuned an octave apart. By themselves, all these features – since they converge – validate each other.

All the examples we have examined show that even if musicians cannot explain in abstract manner the rules followed by the music they perform, they are nevertheless bound by specific and complex constraints. The very existence of these constraints points to an implicit theoretical construct. It is true that theory is implicit, that apprenticeship proceeds mainly through imitation, that verbalisation plays a minor role. Yet, there must be some sort of a musical metalanguage. This raises an important question. How do the bearers of a tradition become aware of the categories which constitute such a metalanguage, and of the terms which sometimes denote them?

While conducting fieldwork among the Aka Pygmies, I once talked to a young woman, an excellent singer by the name of Moako, whom I used to know when she was a little girl. I asked her how she had learned of the existence of those specific names that are used to designate the different vocal parts of a polyphonic ensemble. Specifically, I asked her when and in which circumstances she had heard for the first time the word *ó.sêsê*, which is the name of one of the two parts performed by women. She told me that it happened during a ceremony, when she was a child. All participants, men, women and children, were singing when, suddenly, the *kònzà wà lémbò*, the song leader, told the women to sing the *ó.sêsê*. Immediately one group of women started singing the part that was missing so far. This is how Moako became aware at once of the 'thing' and its name, the sign and its referent. What this example, one among many, tells us is that in the cultures of that region and, most likely, elsewhere as well, the acquisition of the features of musical theory and that of a technical vocabulary occur when mistakes or omissions are noticed, thus through a deductive procedure.

Implicit theories tend to remain latent when everything goes well, a situation often encountered by anthropologists. To the question: 'Why didn't you tell me about that?', the answer will be: 'Well, you didn't ask me . . .'. When no problem arises, terminology remains in the closet, which may have grave consequences. Let us assume, as an hypothesis, that the Aka (or any other ethnic group) made no mistake in their musical practice, and that they were guilty of no omission. Let us also assume that such an idyllic situation lasts for two or three generations. Given the exclusively oral nature of cultural transmission, their musical terminology would then be in danger of disappearing altogether. That this might have actually happened, either among them or in the neighbouring populations, would explain the relative rarity of musical terms in that region today . . . Fortunately, mistakes are only too human. Let us then hope that the local musicians will not fail, once in a while, to bungle their performances.

KNOW-HOW, OR THE USE OF MUSICAL SYSTEMS

In his introduction to *Cognitive Anthropology* (1987), Stephen A. Tyler remarks: 'Obviously, we are interested in the mental codes of other peoples, but how do we infer these mental processes?' He then insists that naming is 'one of the chief methods for imposing order on perception'. He notes however that 'it is probably not true that all named things are significant, just as it is not the case that all significant things are named'. It is indeed quite often the case, in popular traditional music, that significant things are left unnamed. I now want to illustrate the mental representations and the cognitive processes at work in African music in situations which actually seem to involve no naming. For this purpose, I refer to two major musical parameters, pitch and duration.

Pitch is the raw material for scales, that is for the organisation of melodic constructs, whereas duration is the raw material for the organisation of time, that is for rhythmic constructs. A scale is a finite set of discrete units, the degrees. The respective position of the degrees form a contrastive system within the frame of an octave. The diatonic scale, instantiated by the white keys of a piano, has seven degrees. In Central Africa, the most common scale is the pentatonic anhemitonic one (five degrees without half-tones), which roughly corresponds to the black keys of a piano. However, the positions of these degrees may fluctuate between those in a tempered scale (the scale of a pianoforte) and those of an equipentatonic system, in which degrees are separated by intervals of equal size. Thus, different tunings of the same instrument may show a large margin of tolerance, sometimes producing a feeling of uncertainty in the observer: the arrangement of the degrees of the scale may be interpreted in different ways, without thereby altering the pentatonic nature of the ensemble. Therefore, when we listen to a musical piece played on an instrument, it is often extremely difficult to infer the precise modal scale of that instrument from the way it is tuned. In other words, they are ambiguous.

Let us take the case of the xylophone. The scale to which it is tuned can be interpreted in several ways. Furthermore, the sounds it produces are systematically modified by mirlitons added to the calabashes that act as resonators. This produces an effect of 'trompe-l'oreille', making it very difficult to determine the location of the degrees on the scale.

When listening to a piece performed on a xylophone, one first thinks that the instrument is approximately tuned on a scale of the G–A–B–D–E type, with a B and an E slightly too high. On further listening, the B previously perceived as pitched too high will appear as a slightly low C. Listening again, what was considered until then as an E pitched too high is in fact a low F. From G–A–B–D–E, we have come to G–A–C–D–F. The scale seems to change on each new listening, though the tuning of the xylophone remains constant. (The tuning of xylophones varies from one ethnic group to the other.) What happens is that our perception shifts with each hearing. The scale seems uncertain and ambiguous, and one does not know how to interpret it. However, when listening to a choir singing without any melodic accompaniment, the scale seems clear and obvious. But since in this area the instruments and the voices usually perform together, if there is a discrepancy, one has to assume that the tuning of the instruments, albeit different, is indeed intentional. Which means that the tuning obeys another way of conceiving the instrumental scale. What conception? What ideal scale does the maker of the instrument have in mind? In other words, what is the mental template the musicians of a given tradition are able to carry and transmit?

I addressed this problem a few years ago, with my small team at the Centre National de la Recherche Scientifique in Paris. In practical terms, we wanted to understand the principles that controlled the tuning of the several types of xylophones found in this area. In the African context, abstract concepts such as those of 'scale', 'degree' or 'interval' are not verbalised. Within any unverbalised subject, any inquiry based on direct questioning will not work. (I have tried it many times: it just fails.) Therefore, it seemed that the only possible alternative was an experimental procedure. We had to find the tool and the method that would enable local musicians to reply by simply expressing their agreement or disagreement to musical 'stimulations' in the course of natural performance, *in situ*. We wanted to create neutral, that is non-directive, situations.

Indeed, the best thing was to simulate the instrument with a device on which we could modify at will the two basic parameters of sound perception: pitch and timbre. Ideally, various xylophonists would be able to play the music of their respective ethnic groups, using various models of timbre and of musical scales, with the choice of accepting, refusing and also modifying them until satisfied with the result. Such a device would also need a screen to visualise the modifications made, and a memory able to record those changes for immediate study. Fortunately, thanks to the help of several specialists in computer music, we learned that the suitable tool – the only one – did in fact exist, in the unexpected form of a synthesiser, a Yamaha DX7 II FD. With this synthesiser it was possible to combine many operations, among them:

- creating timbres not included in the original pre-programmed repertoire, a necessary condition for the simulation of different xylophones from various ethnic groups
- modifying the order of succession of sounds on the keyboard. This aspect is particularly important because the bars of xylophones of Central Africa are juxtaposed according to an ergonomic principle: their disposition on the frame of the instrument corresponds to the frequency with which they are played, resulting in spatial configurations which are quite different from that of the 'classical' western keyboard
- allowing very fine micro-tuning of each key on the keyboard, so

that each degree can be retuned by the musician himself (in this area xylophones are played by men), until he is satisfied that the device is tuned to his own particular scale

- storing all the data resulting from modifications done by the musicians themselves in its memory

There was a problem, however; namely that keyboards are played with ten fingers but xylophones are hit with two mallets. If people are used to dealing with five to twelve bars, how are they to manage with a keyboard of five octaves? Another difficulty was that the layout of the pitches of a Central African xylophone not only varies from one ethnic group to another, but also from one xylophone to another in the same instrumental ensemble.

We had to transform the synthesiser's keyboard, to 'disguise' it into as many 'bar-boards' as necessary. In other words, we had to simulate the bars of a large variety of xylophones, and also – above all – to allow the players to strike with their mallets exactly as they do on the bars of their own instruments. With this in mind, I had the idea of fixing longish plates of plywood onto the keys, using Velcro tape, which allowed us to rapidly attach or remove the plates or shift their position. We were thus able to simulate at will on the synthesiser as many types of xylophone as necessary, and to program them so that each musician at the DX 7 could imagine he was in front of his own xylophone.

Thanks to these 'removable bars', the fixed-up synthesiser has enabled us to bypass verbalisation by putting an interactive procedure to work. Each of the dozen xylophonists with whom we worked (out of five different ethnic groups) was able: to select the timbre which seemed to him the closest to the one of his own instrument; to play any piece of his repertoire; and to carry out the tuning of his own instrument with a degree of discrimination far beyond the approximate norms used for tuning traditional xylophones in this region.

Supplied with a portable generator, the DX 7 and its series of programmed timbres and scales, we set forth and arrived in the bush. For the first trial run of the new method our choice was the Manza, for in the course of previous field trips, we had worked with the Manza xylophonist Dominique Bawassan, who is an accomplished musician, a man of much intelligence and openness of mind, who had proved a valuable collaborator.

We first tested various previously obtained results. Then we asked the singers whom Bawassan usually accompanied to perform several of their repertoire pieces with him, so that everyone could perfect the memorisation of the timbre and scale of his xylophone. We then invited the musician to come up close to the synthesiser, which had been covered with a piece of cloth, doing this the better to perceive his reactions at the very moment when he discovered this amazing device: a DX 7 transformed into a xylophone!

The moment was crucial. Feeling at once moved and anxious, we were above all afraid of the outright rejection of this device which might appear weird to the musician. Once the synthesiser was unveiled, we explained to him how we had tried to replicate his instrument and suggested he try it out. Straight away, without the slightest hesitation, Dominique seized his mallets and started playing on the synthesiser with stupefying ease. He was almost as quickly joined by the singers.

We soon witnessed an astonishing scene wherein a Manza musician played 'his' xylophone on our synthesiser, manipulating the different controls with ease as he changed from one timbre to another, starting over again, eliminating a timbre, hesitating between the last pair of timbres and finally declaring: 'This one is the twin of my instrument!'

Once the timbre was chosen, it became possible to submit the different pre-programmed scales to the musician's choice. A first discrimination was established between scales accepted and scales declined. The latter were presented again to him so that he might tell us if he thought a scale was incorrect altogether or if only the sounds of certain xylophone bars were not right.

In the last scale configuration, the musician immediately designated one or two bars which he thought had been badly tuned. This let us put another interactive procedure into action: to ask the musician to correct the defective degrees by adjusting the micro-tuning provided on the synthesiser.

Soon, Dominique Bawassan, a traditional xylophonist, was hitting a bar with one hand while using his other hand to adjust the micro-tuning cursor. This was a musician who, just like his ancestors, was used to tune the bars of his xylophone by cutting the wood to proper size (always with the risk of a cut going past the point of no return), who suddenly took his finger off the cursor to tell us, with the utmost simplicity: 'What I've done here is just like when I cut my wood!' The spontaneity of the remark was the evidence that he was in no way perplexed. His gesture was as pertinent in the one case as in the other; only the technical modalities were changed.

It is essential to point out that all the experimental sessions were held in the presence, and with the active participation, of other xylophonists, members of the same ethnic group. Also taking an equal part were the singers used to performing with him. There were also some dignitaries, elders and others locally considered as authorities on the subject. There was practically no disagreement among the different protagonists. That meant that the acceptance or rejection of a particular scale or timbre was the expression of a cultural judgement based on the norms of the collective musical tradition.

We used the same procedure for each xylophone examined. The instruments were sometimes played solo, sometimes in ensembles of three or four. At each new experiment we lived through the same apprehensions: whether our respondents would categorically refuse the experimental device in itself, and then the same emotions when the principle was accepted, and finally the joy of observing that they became involved with as much enthusiasm as patience.

Our synthesiser thus became a discovery device for forms of knowledge and know-how which, by their nature, cannot be verbalised. To put it another way, in revealing what was before implicit, the hand moving the cursor transcended verbal expression. Information stored in the brain of the musician was immediately engraved in the electronic memory. The clear existence of these mental templates, which are specific to each ethnic community, proves that the apparent ambiguity in the tuning of the xylophones is not random. Indeed, it is precisely the principle of tolerance they embody that allows the accompaniment of songs in different pentatonic modes without having to use several xylophones, each tuned to the corresponding mode.

Let us now turn to another essential determinant of the formal structure of music: its temporal organisation. In traditional Africa, the formal organisation of music is always underlaid by rigorous and sophisticated arithmetical principles. These musics are measured; all durations therein are strictly proportional. The musics offer in most cases cyclical structures of the ostinato-with-variations type. In all pieces, periodicity and tempo are fixed: here time is turned in on itself. Any musical utterance occurs within a fixed periodic framework, based on units of time characterised by even integers. These pulsations are regular, that is to say equidistant and of undifferentiated intensity, and they constitute the metric grid of the piece. The 'bar' of Western music, with the strong beat which determines it, is an intermediary between the musical period and the pulsation, a regularly accented matrix. Here there is nothing of the sort. The pulsations that form the period all have equivalent status. A pulsation may be subdivided into two or three very brief durations. These are the minimal operative values: they constitute the smallest pertinent duration, and all other durations are multiples of it.

In a polyphonic and/or polyrhythmic piece, several musical phrases are performed simultaneously. Their various parts do not have the same sizes: different periodicities are thus superimposed. It is important to note, however, that in such contexts the different simultaneously unfolding periods always stand in simple ratios, such as 2:1, 3:1, 3:2, 4:2, and multiples thereof. Whenever two or more periods have no common denominator (as, for example, when they stand in a ratio of 2:3 or 3:4), their beginnings can only coincide at the start of some longer period whose length is necessarily a multiple of the lengths of each. This multiple is thus a macroperiod.

To illustrate the degree of complexity sometimes reached in the music of that region, I will use an Aka Pygmies piece called *Bòbàngì*, from the *yómbè* repertoire (see Figure 3).

This piece requires a polyphonic choir of four voices. Its polyrhythmic substructure is provided by four percussion instruments: three drums and a pair of iron blades. The four vocal parts unfold within a periodic framework of 36 minimal values, while those of the first two drums equal 12 such values and are therefore in a 3:1 ratio to the song period. The period of the third drum is of 3 minimal values. It is therefore in a 4:1 ratio to the periods of the other drums, and in a 12:1 ratio to that of the choir. The cycle of the metal blades lasts 24 minimal values. This is the longest instrumental period. It is in an 1:8 ratio to the third drum and in an 1:2 ratio to the other ones. But as the blades period is of 24 and that of the song of 36, the respective beginnings will only coincide after 3 cycles of this instrument and 2 vocal cycles. What is striking here is that no part displays a cycle of the same length as that of what I have called a macroperiod. Here are the ratios between the various periodicities in *Bòbàngì*:

Between the song and:
- the first two drums 1:3
- the third drum 1:12
- the iron blades 2:3

For the percussion instruments, between:
- the first two drums and the third one 1:4
- the iron blades and the first two drums 1:2
- the iron blades and the third drum 1:8

In short, the organisation of the periodicity of this piece uses six ratios: 1:2, 1:3, 1:4, 1:8, 1:12 and 2:3.

At this level, the piece appears globally symmetrical. A closer look at the way the substance is organised, however, reveals that except for the third drum, the symmetry is systematically broken. And this appears not only in the various types of periodicity present in the piece, but also between parts of identical periodicity. Thus the 36

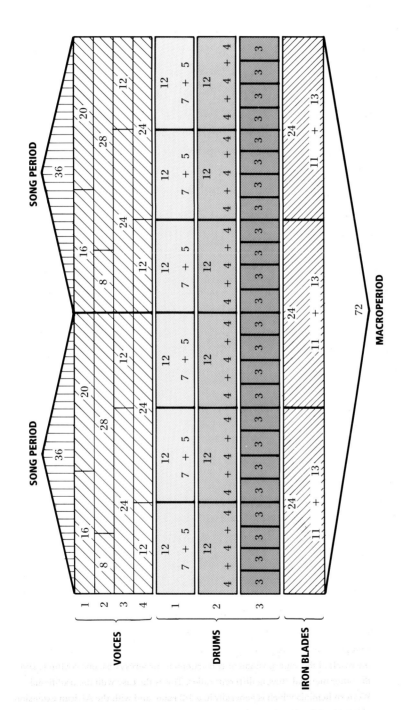

Figure 3 Periodic structure of *Bòbàngì*

minimal values of the vocal period are grouped in segments of 16 + 20 values for the first voice, 8 + 28 for the second, 24 + 12 for the third, and 12 + 24 for the fourth one. The 12-value period shared by the first two drums is in fact divided in very different ways in their respective parts: 7+5 for the first drum, 4 + 4 + 4 for the second one. While the part of the first drum is organised along the principle of rhythmic oddity,[1] that of the second one is a hemiola.[2] As for the 24 minimal values of the blades part, they are grouped in segments of 11 + 13. As does that of the first drum, this part reflects rhythmic oddity.

What is remarkable is that attributing any notion of measuring to the performers is out of the question. Paradoxically, in traditional African music, 'everything is measured but no one counts'. For the African musician, each melodic construct, each rhythmic formula is conceived and perceived as a global, indivisible entity which, once learnt and memorised, will be performed without reference to any counting.

Pieces like *Bòbàngì* confront us with two types of complexity: the first one arises, on the horizontal axis, from the asymmetrical articulation of the different parts, while the second relates to their polyrhythmic entanglement. The result is a set of auditory illusions which create in the listener a feeling of uncertainty or ambiguity. Persisting throughout the performance of the piece, it will only cease with

1 Rhythmic oddity is an extremely subtle property which seems to be found only in African music. It applies to periods for which the total number of pulsations divided by two is an even number. Under this principle, however, any attempt at segmenting the rhythmic content of such periods from whatever starting point will inevitably yield two unequal parts, each composed of an odd number of minimal values. This structure results from an irregular juxtaposition of binary and ternary groups. Such figures are invariably arranged into two blocks containing respectively half + 1 and half −1 of the total minimal values.

2 Hemiola can be defined as the repetition, within a single period, of an identical configuration at different positions with respect to the pulsation. This offsetting is the result of the superposition of two arithmetic progressions, one rhythmic and the other metrical, having different ratios. This is the case with the traditional Western hemiola which is generally in a 3:2 ratio, and with the African extension of it to the 4:3 ratio.

the music itself. This seems to indicate that, as in the melodic organisation of musical scales, ambiguity is a constitutive principle of the rhythmical and polyrhythmical treatment of a large range of traditional African music.

Clearly, as Stephen Tyler says, 'we are interested in the mental codes and cognitive process of the people we work with'. It sometimes happens, however, that in the process these people also learn about us, and come to grasp our own logical procedures. I will illustrate this with a last story, which took place while I was transcribing, in the field, the horn music of the Banda-Linda mentioned above.

I wanted to reconstruct a piece of this repertoire under the supervision of the leader of the orchestra and some musicians he had chosen. I had a series of recordings of instruments coupled two by two: 1–2, 2–3, and so on. Since the combination of the different parts is very complex, I had written down each part on a long strip of paper, so as to be able to match them, two by two. In order to check the accuracy of the different combinations, I intended to sing them and ask the leader to tell me for each combination whether the result was right, and then to assemble the strips in the right position with scotch tape. On the camping-table around which we were all seated were a tape recorder, the long strips of musical paper, and a scotch tape roll. While I was shifting my papers, trying to find the right combination of the first two parts, all were concentrating, observing each of my movements, puzzled by the fact that I could recreate a music, their music, from strips of paper. After having moved the second part under the first one for a while, I judged that the assemblage was correct and started singing the combination of the first two horns on the basis of what I had written down. When I finished singing, and raised my eyes to see the musicians' reaction, nobody said anything except the leader, Djiredji. He watched me and then simply said, very calm, very detached: 'Don't move the paper any more'.

FURTHER READING

Arom, Simha, *African Polyphony and Polyrhythm. Structure and Methodology*,
    Cambridge: Cambridge University Press 1991.
Nketia, J. H. Kwabena, *The Music of Africa*, London: Gollancz 1975.

FURTHER LISTENING

Arom, Simha, *Anthologie de la musique des Pygmées Aka (Centrafrique)*, 2 CD,
    OCORA, C 559012 13. 1987.
Arom, Simha, *Central African Republic*, CD, UNESCO Collection 'Musics and
    Musicians of the World', Auvidis, D 8020 AD 090. 1988.
Arom, Simha, *Banda Polyphonies (Central African Republic)*, CD, UNESCO
    Collection 'Musics and Musicians of the World', Auvidis, D 8043 AD
    090. 1992.
Arom, Simha, *Polyphonies vocales des Pygmées Mbenzele, République
    Centrafricaine*, CD, Inédit Collection, Maison des Cultures du Monde, W
    260 042
    (in cooperation with Denis-Constant Martin). 1992.
Dehoux, Vincent, *Central African Republic. Music for Xylophones*, CD,
    CNRS-Musée de l'Homme Collection, Le Chant du Monde LDX 274 932.
    1992.

FURTHER WATCHING

Arom, Simha, *Polyrhythms in Africa*, Lecture-demonstration with three African
    drummers, video, University of Cambridge Audiovisual Service,
    44 minutes. 1991.

Language and intelligence

DANIEL C. DENNETT

We human beings may not be the most admirable species on the planet, or the most likely to survive for another millennium, but we are without any doubt at all the most intelligent. We are also the only species with language. What is the relation between these two obvious facts? Does thought depend on language?

Before going on to consider that question, I must pause briefly to defend my second premise. Don't whales and dolphins, vervet monkeys and honey bees (the list goes on) have languages of sorts? Haven't chimpanzees in laboratories been taught rudimentary languages of sorts? Yes, and body language is a sort of language, and music is the international language (sort of) and politics is a sort of language, and the complex world of odour and olfaction is another, highly emotionally charged language, and so on. It sometimes seems that the highest praise we can bestow on a phenomenon we are studying is the claim that its complexities entitle it to be called a language – of sorts. This admiration for language – real language, the sort only we human beings use – is well-founded. The expressive, information-encoding properties of real language are practically limitless (in at least some dimensions), and the powers that other species acquire in virtue of their use of proto-languages, hemi-semi-demi-languages, are indeed similar to the powers we acquire thanks to our use of real language. These other species do climb a few steps up the mountain on whose summit we reside, thanks to language. Looking at the vast differences between their gains and ours is one way of approaching the question I want to address.

HOW DOES LANGUAGE CONTRIBUTE TO INTELLIGENCE?

I once saw a cartoon showing two hippopotami basking in a swamp, and one was saying to the other: 'Funny – I keep thinking it's Tuesday!' Surely no hippopotamus could ever think the thought that it is Tuesday. But on the other hand, if a hippopotamus could say that it was thinking any thought, it could probably think the thought that it was Tuesday.

What varieties of thought require language? What varieties of thought (if any) are possible without language? These might be viewed as purely philosophical questions, to be investigated by a systematic logical analysis of the necessary and sufficient conditions for the occurrence of various thoughts in various minds. And in principle such an investigation might work, but in practice it is hopeless. Any such philosophical analysis must be guided at the outset by reflections about what the 'obvious' constraining facts about thought and language are, and these initial intuitions turn out to be treacherous.

We watch a chimpanzee, with her soulful face, her inquisitive eyes and deft fingers, and we very definitely get a sense of the mind within, but the more we watch, the more our picture of her mind swims before our eyes. In some ways she is so human, so insightful, but we soon learn (to our dismay or relief, depending on our hopes) that in other ways, she is so dense, so uncomprehending, so unreachably cut off from our human world. How could a chimp who so obviously understands *A* fail to understand *B*? It sometimes seems flat impossible – as impossible as a person who can do multiplication and division but can't count to ten. But is that really impossible? What about idiot savants who can play the piano but not read music, or children with Williams Syndrome (Infantile Hypercalcemia or IHC) who can carry on hyperfluent, apparently precocious conversations but are so profoundly retarded they cannot clothe themselves?

Philosophical analysis by itself cannot penetrate this thicket of per-plexities. While philosophers who define their terms carefully might succeed in proving logically that – let us say – mathematical thoughts are impossible without mathematical language, such a proof might be consigned to irrelevance by the surprising discovery that math-ematical *intelligence* does not depend on being able to have mathem-atical thoughts so defined!

Consider a few simple quesions about chimpanzees: could chim-panzees learn to tend a fire – could they gather firewood, keep it dry, preserve the coals, break the wood, keep the fire size within proper bounds? And if they couldn't invent these novel activities on their own, could they be trained by human beings to do these things? I wonder. Here is another question. Suppose you imagine something outlandish – say a man climbing up a rope with a plastic dustbin over his head. An easy mental task for you. Could a chimpanzee do the same thing in her mind's eye? I wonder. I chose the elements – man, rope, climbing, dustbin, head – as familiar objects in the perceptual and behavioural world of a laboratory chimp, but I wonder whether a chimp could put them together in this novel way – even by accident, as it were. You were provoked to perform your mental act by my sug-gestion, and probably you often perform similar mental acts on your own in response to suggestions you give yourself – not spoken out loud, but definitely verbalised. Could it be otherwise? Could a chim-panzee get itself to perform such a mental act without the help of verbal suggestion? I wonder.

COMPARING OUR MINDS WITH OTHERS

These are rather simple questions about chimpanzees, but neither you nor I know the answers – yet. The answers are not impossible to acquire, but not easy either; controlled experiments could yield the answers, which would shed light on the role of language in turning brains into minds like ours. I think it is very likely that every content that has so far passed through your mind while reading this chapter is strictly off limits to non-language users, be they apes or dolphins, or even non-signing deaf people. If this is true, it is a striking fact, so striking that it reverses the burden of proof in what otherwise would be a compelling argument: the claim, first advanced by the linguist Noam Chomsky, and more recently defended by the philosophers Jerry Fodor and Colin McGinn, that our minds, like those of all other species, must suffer 'cognitive closure' with regard to some topics of inquiry. Spiders can't contemplate the concept of fishing, and birds – some of whom are excellent at fishing – are not up to thinking about democracy. What is inaccessible to the dog or the dolphin may be readily grasped by the chimp, but the chimp in turn will be cognitively closed to some domains we human beings have no difficulty thinking about. Chomsky and company ask a rhetorical question: What makes us think we are different? Aren't there bound to be strict limits on what *Homo sapiens* may conceive? This presents itself as a biological, naturalistic argument, reminding us of our kinship with the other beasts, and warning us not to fall into the ancient trap of thinking 'how like an angel' we human 'souls', with our 'infinite' minds are.

I think that on the contrary it is a pseudo-biological argument, one that by ignoring the actual biological details, misdirects us away from the case that can be made for taking one species – our species – right off the scale of intelligence that ranks the pig above the lizard and the ant above the oyster. Comparing our brains with bird brains or dolphin brains is almost beside the point, because our brains are in effect joined together into a single cognitive system that dwarfs all others. They are joined by one of the innovations that has invaded our brains and no others: language. I am not making the foolish claim that all our brains are knit together by language into one gigantic mind, thinking its transnational thoughts, but rather that each individual human brain, thanks to its communicative links, is the beneficiary of the cognitive labours of the others in a way that gives it unprecedented powers. Naked animal brains are no match at all for the heavily armed and outfitted brains we carry in our heads.

A purely philosophical approach to these issues is hopeless. It must be supplemented – not replaced – with researches in a variety of disciplines ranging from cognitive psychology and neuroscience to evolutionary theory and palaeoanthropology. I raised the question

about whether chimps could learn to tend a fire because of its close –
but treacherous! – resemblance to questions that have been discussed
in the recent flood of excellent books and articles about the evolution
of the human mind.

I will not attempt to answer the big questions, but simply explain
why answers to them will hinge on answers to the questions raised –
and to some degree answered – in this literature. In the terms of the
Oxford zoologist Richard Dawkins, my role today is to be a vector
of memes, attempting to infect the minds in one niche – my home
discipline of philosophy – with memes that are already flourishing in
others.

At some point in prehistory, our ancestors tamed fire; the evidence
strongly suggests that this happened hundreds of thousands of
years – or even as much as a million years – before the advent of language, but of course after our hominid line split away from the
ancestors of modern apes such as chimpanzees. What, if not language, gave the first fire-taming hominids the cognitive power to
master such a project? Or is fire-tending not such a big deal? Perhaps
the only reason we don't find chimps in the wild sitting around
campfires is that their rainy habitats have never left enough tinder
around to give fire a chance to be tamed. (The neurobiologist William
Calvin tells me that Sue Savage-Rumbaugh's pygmy chimps in
Atlanta love to go on picnics in the woods, and enjoy staring into the
campfire's flames, just as we do.)

NEED TO KNOW VERSUS THE COMMANDO TEAM

If termites can create elaborate, well-ventilated cities of mud, and
weaverbirds can weave audaciously engineered hanging nests, and
beavers can build dams that take months to complete, couldn't chimpanzees tend a simple campfire? This rhetorical question climbs
another misleading ladder of abilities. It ignores the independently
well-evidenced possibility that there are two profoundly different
ways of building dams: the way beavers do and the way we do. The

differences are not necessarily in the products, but in the control structures within the brains that create them. A child might study a weaverbird building its nest, and then replicate the nest herself, finding the right pieces of grass, and weaving them in the right order, creating, by the very same series of steps, an identical nest. A film of the two building processes occurring side-by-side might overwhelm us with a sense that we were seeing the same phenomenon twice, but it would be a mistake to impute to the bird the sort of thought processes we know or imagine to be going on in the child. There could be very little in common between the processes going on in the child's brain and the bird's brain. The bird is (apparently) endowed with a collection of interlocking special-purpose minimalist subroutines, well-designed by evolution according to the notorious 'need to know principle' of espionage: give each agent as little information as will suffice for it to accomplish its share of the mission.

Control systems designed under this principle can be astonishingly successful – witness the birds' nests, after all – whenever the environment has enough simplicity and regularity, and hence predictability, to favour predesign of the whole system. The system's very design in effect makes a prediction – a wager, in fact – that the environment will be the way it must be for the system to work. When the complexity of encountered environments rises, however, and unpredictability becomes a more severe problem, a different design principle kicks in: the commando team principle illustrated by such films as *The Guns of Navarone* – give each agent as much knowledge about the total project as possible, so that the team has a chance of ad libbing appropriately when unanticipated obstacles arise.

Fortunately, we don't have to inspect brain processes directly to get evidence of the degree to which one design principle or the other is operating in a particular organism – although in due course it will be wonderful to get confirmation from neuroscience. In the meantime, we can conduct experiments that reveal the hidden dissimilarities by showing how bird and child respond to abnormal obstacles and opportunities along the way.

I have a favourite example of such an experiment with beavers. It turns out that beavers hate the sound of running water and will cast about frantically for something – anything – that will bring relief. As Wilsson has shown, when recordings of running water are played from loudspeakers, beavers will respond by plastering the loud-speakers with mud.

So there is a watershed in the terrain of evolutionary design space; when a control problem lies athwart it, it could be a matter of chance which direction evolution propelled the successful descendants. Perhaps, then, there are two ways of tending fires – roughly, the beaver-dam way, and our way. If so, it is a good thing for us that our ancestors did not hit upon the beaver-dam way. If they had, the woods might today be full of apes sitting around campfires, but we would not be here to marvel at them.

THE TOWER OF GENERATE-AND-TEST

I want to propose a framework in which we can place the various design options for brains, to see where their power comes from. It is an outrageously oversimplified structure, but idealisation is the price one should often be willing to pay for synoptic insight. I will call it the tower of generate-and-test.

In the beginning there was Darwinian evolution of species by natural selection. A variety of candidate organisms were blindly generated by more or less arbitrary processes of recombination and mutation of genes. These organisms were field tested, and only the best designs survived. This is the ground floor of the tower. Let us call its inhabitants *Darwinian creatures*. (Is there perhaps a basement? Recently speculations by physicists and cosmologists about the evolution of universes open the door to such a prospect, but I will not explore it on this occasion. My topic today is the highest stories of the tower.)

This process went through many millions of cycles, producing many wonderful designs, both plant and animal, and eventually among its novel creations were some designs with the property of phenotypic plasticity. The individual candidate organisms were not wholly designed at birth, or in other words there were elements of their design that could be adjusted by events that occurred during the field tests. Some of these candidates, we may suppose, were no better off than their hard-wired cousins, since they had no way of favouring (selecting for an encore) the behavioural options they were equipped to 'try out', but others, we may suppose, were fortunate enough to have wired-in 'reinforcers' that happened to favour smart moves, actions that were better for their agents. These individuals thus confronted the environment by generating a variety of actions, which they tried out, one by one, until they found one that 'worked'. We may call this subset of Darwinian creatures, the creatures with conditionable plasticity, *Skinnerian creatures*, since, as B. F. Skinner was fond of pointing out, operant conditioning is not just analogous to Darwinian natural selection; it is continuous with it. 'Where inherited behavior leaves off, the inherited modifiability of the process of conditioning takes over.'

Skinnerian conditioning is a fine capacity to have, so long as you are not killed by one of your early errors. A better system involves preselection among all the possible behaviours or actions, weeding out the truly stupid options before risking them in the harsh world. We human beings are creatures capable of this third refinement, but we are probably not alone. We may call the beneficiaries of this third story in the tower *Popperian* creatures, since as Sir Karl Popper once elegantly put it, this design enhancement 'permits our hypotheses to die in our stead.' Unlike the merely Skinnerian creatures who survive because they are lucky, we Popperian creatures survive because we are smart – of course we are just lucky to be smart, but that is better than just being lucky.

But how is this preselection in Popperian agents to be done? Where is the feedback to come from? It must come from a sort of inner environment – an inner something-or-other that is structured in such a way that the surrogate actions it favours are more often than not the very actions the real world would also bless, if they were actually performed. In short, the inner environment, whatever it is, must contain lots of information about the outer environment and its regularities.

Nothing else (except magic) could provide preselection worth having. Now here we must be very careful not to think of this inner environment as simply a replica of the outer world, with all its physical contingencies reproduced. (In such a miraculous toy world, the little hot stove in your head would be hot enough to actually burn the little finger in your head that you placed on it!) The information about the world has to be there, but it also has to be structured in such a way that there is a non-miraculous explanation of how it got there, how it is maintained, and how it actually achieves the preselective effects that are its *raison d'être*.

We have now reached the story of the Tower on which I want to build. Once we get to Popperian creatures, creatures whose brains have the potential to be shaped into inner environments with preselective prowess, what happens next? How does new information about the outer environment get incorporated into these brains? This is where earlier design decisions – and in particular, choices between need to know and commando team – come back to haunt the designer; for if a particular species' brain design has already gone down the need to know path with regard to some control problem, only minor modifications (fine tuning, you might say) can be readily made to the existing structures, so the only hope of making a major revision of the internal environment to account for new problems, new features of the external environment that matter, is to submerge the old hard-wiring under a new layer of pre-emptive control (a theme developed in the work of the Artificial Intelligence (AI) researcher Rodney Brooks). It is these higher levels of control that have the potential for vast increases in versatility. And it is at these levels, in particular, that we should look for the role of language (when it finally arrives on the scene), in turning our brains into virtuoso preselectors.

We engage in our share of rather mindless routine behaviour, but our important acts are often directed on the world with incredible cunning, composing projects exquisitely designed under the influence of vast libraries of information about the world. The instinctual actions we share with other species show the benefits derived by the harrowing explorations of our ancestors. The imitative actions we share with some higher animals may show the benefits of information gathered not just by our ancestors, but also by our social groups over generations, transmitted non-genetically by a 'tradition' of imitation. But our more deliberately planned acts show the benefits of information gathered and transmitted by our conspecifics in every culture, including, moreover, items of information that no single individual has embodied or understood in any sense. And while some of this information may be of rather ancient acquisition, much of it is brand new. When comparing the timescales of genetic and cultural evolution, it is useful to bear in mind that we today – every one of us – can easily understand many ideas that were simply unthinkable *by the geniuses* in our grandparents' generation!

The successors to mere Popperian creatures are those whose inner environments are informed by the designed portions of the outer environment. We may call this sub-sub-subset of Darwinian creatures *Gregorian creatures*, since Richard Gregory is to my mind the preeminent theorist of the role of information – or more exactly, what Gregory calls potential intelligence – in the creation of smart moves – or what Gregory calls kinetic intelligence. Gregory observes that a pair of scissors, as a well-designed artifact, is not just a result of intelligence, but an endower of intelligence (external potential intelligence), in a very straightforward and intuitive sense: when you give someone a pair of scissors, you enhance their potential to arrive more safely and swiftly at smart moves.

Anthropologists have long recognised that the advent of tool use accompanied a major increase in intelligence. Our fascination with the discovery that chimpanzees in the wild fish for termites with crudely prepared fishing sticks is not misplaced. This fact takes on further significance when we learn that not all chimpanzees have hit upon this trick; in some chimpanzee 'cultures' termites are a present but unexploited food source. This reminds us that tool use is a two-way sign of intelligence; not only does it require intelligence to recognise and maintain a tool (let alone fabricate one), but it confers intelligence on those who are lucky enough to be given the tool. The better designed the tool, the more information is embedded in its fabrication, the more potential intelligence it confers on its user. And among the pre-eminent tools, Gregory reminds us, are what he calls mind-tools: words. What happens to a human or hominid brain when it becomes equipped with words? I have arrived, finally, back at the question with which I began.

WHAT WORDS DO TO US

There are two related mistakes that are perennially tempting to theorists thinking about the evolution of language and thinking. The first is to suppose that the manifest benefits of communication to humanity (the group, or the species) might themselves explain the evolution of language. The default supposition of evolutionary theory must be that individuals are initially competitive, not cooperative, and while this default can be most interestingly overridden by special conditions, the burden is always to demonstrate the existence of the special conditions. The second mistake is to suppose that mind-tools – words, ideas, techniques – that were not 'good for us' would not survive the competition. The best general antidote I know to both these errors is Richard Dawkins' discussion of *memes* in *The Selfish Gene*. The best detailed discussion I know of the problem of designing communication under the constraint of competitive communicators is by Dan Sperber and Deirdre Wilson in their excellent book, *Relevance: A Theory of Communication*.

One upshot of the considerations raised by these thinkers is that one may usefully think of words – the most effective vehicles for memes – as invading or parasitising a brain, not simply being acquired by a brain. What is the shape of this environment when words first enter it? It is definitely not an even playing field or a *tabula rasa*. Our newfound words must anchor themselves on the hills and valleys of a landscape of considerable complexity. Thanks to earlier evolutionary pressures, our innate quality spaces are species-specific, narcissistic, and even idiosyncratic from individual to individual.

A number of investigators are currently exploring portions of this terrain. The psychologist Frank Keil and his colleagues at Cornell have evidence that certain highly abstract concepts – such as the concepts of being alive or ownership, for instance – have a genetically imposed head start in the young child's kit of mind-tools; when the specific words for owning, giving and taking, keeping and hiding, and their kin enter a child's brain, they find homes already partially built for them. Ray Jackendoff and other linguists have identified fundamental structures of spatial representation – notably designed to enhance the control of locomotion and the placement of movable things – that underlie our intuitions about concepts like *beside*, *on*, *behind*, and their kin. Nicholas Humphrey has argued in recent years that there must be a genetic predisposition for adopting what I have called the intentional stance, and Alan Leslie and others have developed evidence for this, in the form of what he calls a 'theory of mind module' designed to generate second-order beliefs (beliefs about the beliefs and other mental states of others). Some autistic children seem to be well-described as suffering from the disabling of this module, for which they can occasionally make interesting compensatory adjustments.

We are only just beginning to discern the details of the interactions between such pre-existing information structures and the arrival of language, so theorists who have opportunistically ignored the phenomenon up till now have nothing to apologise for. The time has come, however, to change tactics. In Artificial Intelligence, for instance, even the most ambitiously realistic systems – such as Soar, the star of Allen Newell's *Unified Theories of Cognition* – are described without so much as a hint about which features, if any, are dependent on the system's having acquired a natural language with which to supplement its native representational facilities. The result is that most AI agents, the robotic as well as the bedridden, are designed on the model of the walking encyclopedia, as if all the information in the inner environment were in the form of facts told at one time or another to the system. And in the philosophy of mind, there is a similar tradition of theory-construction and debate about the nature of belief, desire and intention – philosophical 'theories of mental representation' – fed on a diet exclusively drawn from language-infected cognitive states. Tom believes that snow is white. Do polar bears believe that snow is white? In the same sense? Supposing one might develop a good general theory of belief by looking exclusively at such specialised examples is like supposing one might develop a good general theory of motor control by looking exclusively at examples of people driving automobiles in city traffic. 'Hey, if that isn't *motor* control, what is?' – a silly pun echoed, I am claiming, by the philosopher who says 'Tom believes snow is white – hey, if that isn't a belief, what is?'

WHAT WORDS DO FOR US

John Holland, a pioneer researcher on genetic algorithms, has recently summarised the powers of the Popperian internal environment, adding a nice wrinkle:

> An internal model allows a system to look ahead to the future consequences of current actions, without actually committing itself to those actions. In particular, the system can avoid acts that would set it irretrievably down some road to future disaster ('stepping off a cliff'). Less dramatically, but equally important, the model enables the agent to make current 'stage-setting' moves that set up later moves that are obviously advantageous. The very essence of a competitive advantage, whether it be in chess or economics, is the discovery and execution of stage-setting moves.

But how intricate and long-range can the 'stage-setting' look-ahead be without the intervention of language to help control the manipulation of the model? This is the relevance of my question at the outset about the chimp's capacities to visualise a novel scene. As Merlin Donald points out, Darwin was convinced that language was the prerequisite for 'long trains of thought', and this claim has been differently argued for by several recent theorists, especially Julian Jaynes and Howard Margolis. Long trains of thought have to be controlled, or they will wander off into delicious if futile woolgathering. These authors suggest, plausibly, that the self-exhortations and reminders made possible by language are actually essential to maintaining the sorts of long-term projects only we human beings engage in (unless, like the beaver, we have a built-in specialist for completing a particular long-term project).

Merlin Donald resists this plausible conjecture, and offers a variety of grounds for believing that the sorts of thinking that we can engage in without language are remarkably sophisticated. I commend his argument to your attention in spite of the doubts about it I will now briefly raise. Donald's argument depends heavily on two sources of information, both problematic in my opinion. First, he makes strong claims about the capabilities of those congenitally deaf human beings who have not yet developed (so far as anyone can tell) any natural language – in particular, signing.

Second, he draws our attention to the amazing case of Brother John, a French Canadian monk who suffers from frequent epileptic seizures that do not render him unconscious or immobile, but just totally aphasic, for periods of a few minutes or hours. During these paroxysms of aphasia, we are told, Brother John has no language, either external or internal. That is, he can neither comprehend nor produce words of his native tongue, not even 'to himself'. At the same time, Brother John can 'still record the episodes of life, assess events, assign meanings and thematic roles to agents in various situations, acquire and execute complex skills, learn and remember how to behave in a variety of settings.'

My doubts about the use to which Donald wants to put these findings are straightforward, and should be readily resolvable in time: both Brother John and the long-term language-less deaf people, are in different ways and to different degrees, still the beneficiaries of the shaping role of language. In the case of Brother John, his performance during aphasic paroxysm relies on 'language-mediated apprenticeships'.

> Brother John maintains, for instance, that he need not tell himself the words 'tape recorder', 'magnetic tape', 'red button on the left', 'turn', 'push' and so forth . . . in order to be capable of properly operating a tape recorder . . . .

The deaf who lack sign – a group whose numbers are diminishing today, thank goodness – lack Brother John's specific language-mediated apprenticeships, but we simply don't know – yet – what structures in their brains are indirect products of the language that most of their ancestors in recent millennia have shared. The evidence that Donald adduces for the powers of language-less thought is thus potentially misleading. These varieties of language-less thought, like barefoot waterskiing, may be possible only for brief periods, and only after a preparatory period that includes the very feature whose absence is later so striking.

There are indirect ways of testing the hypotheses implied by these doubts. Consider episodic memory, for instance. When a dog retrieves a bone it has buried, it manifests an effect on its memory, but must the dog, in retrieving the bone, actually recollect the episode of burying? (Perhaps you can name the current US Secretary of State, but can you recall the occasion of learning his name?) The capacity for genuine episodic recollecting – as opposed to semantic memory installed by a single episode of learning – is in need of careful analysis and investigation. Donald follows Jane Goodall in claiming that chimpanzees in the wild are 'able to perceive social events accurately and to remember them' – as episodes in memory. But we have not really been given any evidence from which this strong thesis follows; the social perspicuity of the chimpanzees might be largely due to specialised perceptual talents interacting with specialised signs. Suppose, for instance, that there is something subtle about the posture of a subordinate facing a superior that instantly – visually – tells an observer chimp (but not a human observer) which is subordinate, and how much. Experiments that would demonstrate a genuine capacity for episodic memory in chimpanzees would have to involve circumstances in which an episode was observed or experienced, but in which its relevance as a premise for some social inference was not yet determined – so no 'inference' could be drawn at once. If something that transpired later suddenly gave a retrospective relevance to the earlier episode, and if a chimpanzee can tumble to that fact, this would be evidence – but not yet conclusive evidence – of episodic memory.

Another way of testing for episodic memory in the absence of language would be to let a chimpanzee observe – once – a relatively novel and elaborate behavioural sequence that accomplishes some end (for example, to make the door open, you stamp three times, turn in a circle and then push both buttons at once), and see if the chimpanzee, faced with the need to accomplish the same end, can even come close to reproducing the sequence. It is not that there is any doubt that chimpanzee brain tissue is capable of storing this much information – it can obviously store vastly more than is required for such a simple feat – but whether the chimpanzee can exploit this storage medium in such an adaptive way on short notice. And that is the sort of question that no amount of microscopic brain-study is going to shed much light on.

THE ART OF MAKING MISTAKES

This brings me to my final step up the tower of generate-and-test. There is one more embodiment of this wonderful idea, and it is the one that gives our minds their greatest power: once we have language – a bountiful kit of mind-tools – we can use it in the structure of deliberate, foresightful generate-and-test known as science. All the other varieties of generate-and-test are willy-nilly.

The soliloquy that accompanies the errors committed by the lowliest Skinnerian creature might be 'Well, I mustn't do *that* again!' and the hardest lesson for any agent to learn, apparently, is how to learn from one's own mistakes. In order to learn from them, one has to be able to contemplate them, and this is no small matter. Life rushes on, and unless one has developed positive strategies for recording one's tracks, the task known in AI as credit assignment (also known, of course, as blame assignment!) is insoluble. The advent of high-speed still photography was a revolutionary technological advance for science because it permitted human beings, for the first time, to examine complicated temporal phenomena not in real time, but in their own good time – in leisurely, methodical backtracking analysis of the traces they had created of those complicated events. Here a technological advance carried in its wake a huge enhancement in cognitive power. The advent of language was an exactly parallel boon for human beings, a technology that created a whole new class of objects-to-contemplate, verbally embodied surrogates that could be reviewed in any order at any pace. And this opened up a new dimension of self-improvement – all one had to do was to learn to savour one's own mistakes.

But science is not just a matter of making mistakes, but of making mistakes in public. Making mistakes for all to see, in the hopes of getting the others to help with the corrections. It has been plausibly maintained, by Nicholas Humphrey, David Premack and others, that chimpanzees are natural psychologists – what I would call second-order intentional systems – but if they are, they nevertheless lack a crucial feature shared by all human natural psychologists, folk and professional varieties: they never get to compare notes. They never dispute over attributions, and ask for the grounds for each others' conclusions. No wonder their comprehension is so limited. Ours would be, too, if we had to generate it all on our own.

Let me sum up the results of my rather swift and superficial survey. Our human brains, and only human brains, have been armed by habits and methods, mind-tools and information, drawn from millions of other brains to which we are not closely related. This, amplified by the deliberate use of generate-and-test in science, puts our minds on a different plane from the minds of our nearest relatives among the animals. This species-specific process of enhancement has become so swift and powerful that a single generation of its design improvements can now dwarf the R-and-D efforts of millions of years of evolution by natural selection. So while we cannot rule out the possibility in principle that our minds will be cognitively closed to some domain or other, no good 'naturalistic' reason to believe this can be discovered in our animal origins. On the contrary, a proper application of Darwinian thinking suggests that, if we survive our current self-induced environmental crises, our capacity to comprehend will continue to grow by increments that are now incomprehensible to us.

FURTHER READING

Calvin, William, *The Ascent of Mind: Ice Age Climates and the Evolution of Intelligence*, New York: Bantam 1990.

Dawkins, Richard, *The Selfish Gene*, Oxford: Oxford University Press 1976.

Diamond, Jared, *The Third Chimpanzee: The Evolution and Future of the Human Animal*, New York: Harper 1992.

Donald, Merlin, *Origins of the Modern Mind: Three Stages in the Evolution of Culture and Cognition*, Cambridge, MA: Harvard University Press 1991.

Gregory, Richard, *Mind in Science*, Cambridge: Cambridge University Press 1981.

Jackendoff, Ray, *Consciousness and the Computational Mind*, Cambridge, MA: MIT Press 1987.

Jaynes, Julian, *The Origins of Consciousness in the Breakdown of the Bicameral Mind*, Boston: Houghton Mifflin 1976.

McGinn, Colin, *The Problem of Consciousness*, Oxford: Blackwell 1990.

Newell, Allen, *Unified Theories of Cognition*, Cambridge, MA: Harvard University Press 1990.

Margolis, Howard, *Patterns, Thinking and Cognition*, Chicago: University of Chicago Press 1987.

Humphrey, Nicholas, *The Inner Eye*, London: Faber and Faber 1986.

Premack, David, *Gavagai! or the Future History of the Animal Language Controversy*, Cambridge, MA: MIT Press 1986.

Skinner, B. F., *Science and Human Behavior*, New York: Macmillan 1953.

Sperber, Dan and Wilson, Deirdre, *Relevance: A Theory of Communication*, Cambridge, MA: Harvard University Press 1986.

Wilsson, L., 'Observations and Experiments on the Ethology of the European Beaver', *Viltrevy, Swedish Wildlife*, 8 (1974) 115–266.

Understanding verbal understanding

DAN SPERBER

All of us humans speak and understand at least one language, English for instance. How intelligent we are! Well, how intelligent are we? Is mastery of a language truly proof of intelligence?

I take it that intelligence, like beauty, is a property rather than a thing. There is no area of a brain that might properly be called its intelligence. On the other hand, a variety of doings and doers can be called intelligent. Though there probably isn't a satisfactory definition, typical instances of intelligent doings are not too hard to characterise. They involve creative reasoning achieved by bringing together various information (for instance new and old information) on the basis of which novel conclusions, insights, or decisions can be reached. More technically, we happily describe as intelligent inferential processes that are not entirely 'data-driven'. If a process is not at all inferential, or if it is an automatic reaction to a specific stimulus, then talk of intelligence is much less appropriate.

Notwithstanding the doubts that surround the very notion of intelligence, the view that human language is proof of human intelligence may well be the oldest and most common philosophical cliché. Here is a typical quote:

> As the soft lips and pliant tongue are taught
> With other minds to interchange the thought;
> And sound, the symbol of the sense, explains
> In parted links the long ideal trains;

From clear conceptions of external things
The facile power of Recollection springs.
Whence REASON's empire o'er the world presides
And man from brute, and man from man divides

ERASMUS DARWIN, *The Temple of Nature* 4. 265–72

The claim that linguistic behaviour shows human superior intelligence can be taken in three ways. It may be the things we say: our words give evidence of our thoughts, and the soundness and creativity of our thoughts show how intelligent we are.

It may also be argued that, without linguistic communication, there would be no literature, no science, no law, in a nutshell, no cumulative building of knowledge, theoretical or practical. Now the cultural transmission of knowledge, the Darwin Lectures for instance, no doubt contributes greatly to the development of individual intelligence. These two lines of argument have been soundly and creatively pursued by Daniel Dennett in his contribution to this volume, and I won't consider them further here.

According to a third line of argument, nothing better establishes our intelligence than the very existence of verbal communication, whatever we say. As Descartes argued, even fools saying foolish things display a form of intelligence that no other animal possesses. The very ambiguity of the word 'understanding', meaning both intelligence and comprehension, is, in this respect, quite telling. This view of the relationship between language and intelligence, however, does not square too well with an even more common view, according to which verbal communication is a matter of coding and decoding. For are coding and decoding truly intelligent activities?

CODING AND ITS LIMITS

What do we do when we communicate? What cognitive skill do we thereby exhibit? The common view is that communication is possible just in case interlocutors share a code. A language such as English is seen as a complex code. Whether simple or complex, a code is a device that generates pairs made up of a message and a signal: the Morse code, for instance pairs each letter of the alphabet with a series of short or long beeps; a language pairs linguistic senses and sounds. The pairing of messages and signals generated by the code can be made to work in two kinds of devices: encoders and decoders.

Humans can perform both as encoders of linguistic senses and as decoders of linguistic sound, and that, so the story goes, is how they communicate with one another. Failures of communication occur when encoding or decoding is not done properly, or when noise damages the sound signal, or, more significantly, when the codes of the interlocutors are not properly matched. Otherwise, such code-based communication is sure to run smoothly. This is a simple and powerful explanation of the successes and failures of communication. However, if this explanation is correct, then the ability to communicate linguistically shouldn't be described as intelligent at all.

The work of an encoding or decoding device is neither inferential nor creative. It is not inferential because the symmetrical relation between a message and a signal is quite different from the asymmetrical relation of premise to conclusion: just as the letter 'm' does not

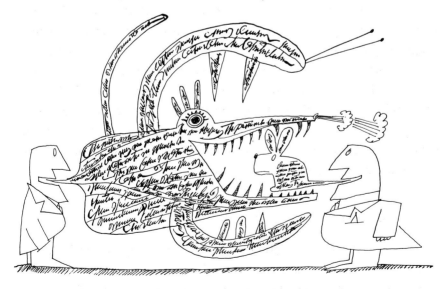

Figure 1 What is communicated is never wholly encoded. Cartoon by Saul Steinberg

logically follow from two long beeps, the meaning of a sentence does not logically follow from its sound. The work of an encoding or decoding device is not creative either: it is an automatic reaction to the input message or signal. Actually, it had better *not* be creative: a creative bout at the encoding or decoding end would jeopardise the symmetry between the two processes and hence the success of code-based communication.

However, the common view of verbal communication is false and, as I will try to show, coding and decoding are just ancillary components of what is essentially a creative inferential process.

Take an ordinary sentence such as 'It's late.' As an English speaker, you know what this means. In a situation where you wanted to convey that meaning to another English speaker, you would say 'It's late' and be understood, and that is all there is to it. Or is it?

To begin with, in saying 'It's late' you might intend to convey not just the explicit information that it is late, but also something implicit: say, that it is time to go home. The existence of such implicit content has long been recognised, and its study, inspired by the work of the philosopher Paul Grice, has become the main focus of pragmatics. Not everything that is communicated is wholly encoded: that much is generally agreed. I would like to make a stronger claim: nothing that is communicated is wholly encoded. In other words, even the explicit part of communication is never fully explicit, not by a wide margin, actually.

Consider the following dialogue:

Peter: When does the train arrive?
Mary: It's late!

Here Mary is using 'It's late' in a meaning different from the one you probably had in mind a minute ago. Whereas in the earlier utterance, the 'it' did not refer to anything at all and was just a syntactic filler, in Mary's utterance, the 'it' refers to the train. In yet other utterances of the same sentence, 'it' might refer to an apology, a payment, a chemical reaction, or whatnot. Analysing the word 'it' in the hope of finding out what, if anything, it refers to would be as unhelpful as staring at a pointed finger in the hope of finding out what, if anything, it is pointing at. What must be taken into consideration in order to determine what 'it' may refer to is, of course, the context.

An utterance must be taken together with a context. How is that done? It is often supposed that there must exist some system of rules that applies to an utterance and its context taken together, to yield the intended interpretation. However, this presupposes that the context is somehow given and, together with the utterance itself, provides well circumscribed initial data for the interpretation process. This presupposition is quite mistaken.

'It's late', taken without contextual information, suggests that the 'it' is a syntactic filler without a referent. Knowing just that 'It's late' was a reply to the question: 'When does the train arrive?' suggests that 'it' referred to the train. But further contextual information might reverse this interpretation:

> *Mary*: I'm tired. I want to go home. We don't even know that Johnny will be on the train.
> *Peter*: When does the train arrive?
> *Mary*: It's late! Forget the train! Let's go home!

Here, the 'it' in 'It's late' is once more the usual syntactic filler. Further contextualisations would again reverse the preferred interpretation. As such examples show, the context is not a given on the basis of which comprehension might proceed. Rather, deciding what constitutes the pertinent context is part and parcel of the interpretation process.

Whether 'it' refers, and to what, is by no means the only source of indeterminacy in 'It's late'. Nothing is late in and of itself. To be late is to be late with respect to some expectation, schedule or timetable. But, of course, several such schedules may enter into consideration. Even referring to specific objects such as trains doesn't determine a single relevant schedule. Thus, I might say, 'The late train to Cambridge is hardly ever late', without contradicting myself: my two uses of 'late' would merely relate to two different schedules: lateness in the day and lateness on departure or on arrival. Deciding which schedule is intended by the speaker again depends on a context that must be discovered as part of the interpretation process.

Suppose, however, that we know that 'it' refers to the train to Cambridge, and that the train is said to be late with respect to its expected time of arrival, 5.25 p.m. Surely, in that case, we know what 'It's late' means. Or do we? Here we are, let us imagine, on the platform at Cambridge railway station at 5.20. Three people are about to utter the same sentence, 'It's late', referring to the same train and with the same schedule in mind and yet they will mean quite different things.

First Peter, who believes the worst of British Rail, asserts at 5.21, 'No doubt it's late again!' However at 5.24 the train is heard, and Mary says to Peter, 'You're a real seer! Indeed it's late!' Our third character is a railway controller looking at his stopwatch. He sees that the train stops at 5.26, and says sternly, 'It's late!' Peter and the controller are speaking in earnest, while Mary is speaking ironically. Peter and Mary use 'late' rather loosely, while the controller uses it so strictly as to see sixty seconds late as late enough to mention. Because of these subtle differences, their three utterances, though the words, the train, and the schedule are the same, convey quite different thoughts.

As an English speaker, you know the linguistic sense, or rather senses, of 'It's late', but these senses are very incomplete affairs; they are not, by themselves, the meanings you might want to convey when you utter the words. As a hearer too, merely knowing the linguistic senses of 'It's late' does not tell you what a speaker means in uttering these words. The same is true of every sentence in every human language: the thought we intend to convey can never be fully encoded, and linguistic decoding is only a first step in understanding a speaker's meaning.

Going from linguistic sense to speaker's meaning is not a matter of further decoding, not even of context-sensitive decoding: it is a matter of inference. What then is the form of inference involved in verbal understanding?

INFERENCE

In any instance of verbal understanding, there is an initial premise and a goal. The premise is the information that a certain person uttered a certain sentence. The goal is to discover what that person meant in uttering that sentence. The premise and the conclusion of a successful comprehension process are both complex in the same way, but not to the same degree.

The first premise in comprehension might be something like this:

Carol says: 'It's late'

This conclusion might be something like this:

Carol means that it is time to go home

Here, premise and conclusion both involve a 'meta-representation': that is, they contain a representation of a representation. The premise is about an utterance ('It's late') and directly quotes that utterance. The conclusion is about a thought (that it is time to go home) and indirectly quotes that thought. On closer analysis, however, it turns out that while the premise is a first-order meta-representation, the conclusion is a higher-order meta-representation.

Just like the attribution of an utterance, the attribution of a belief, for instance

Carol believes that it is time to go home,

may be a simple meta-representational affair. There is an important relationship between meaning and believing but it is not a simple one.

When John concludes that, in saying 'It's late', Carol means that it is time to go home, he need not at all attribute to her the belief that it is time to go home: after all, she might be insincere and attempting to communicate what she does not believe. What John must be attributing to Carol is an intention rather than a belief: the intention that *he* should believe that it is time to go home.

An intention is a mental representation of a desired state of affairs. What state of affairs does a speaker desire? A state of affairs in which some information becomes represented in her hearer's mind as a result of her utterance. We call such intentions 'informative intentions'. Simple informative intentions are first-order meta-representations. Thus the content of Carol's intention is the following meta-representation (I will use different lines and indents to separate the different representational levels):

> John should believe
> > that it is time to go home

Thinking that someone has an informative intention is entertaining a second-order meta-representation. Thus John's understanding of Carol's utterance contains the following second-order meta-representation:

> She intends
> > me to believe
> > > that it is time to go home

I will argue later that full-fledged comprehension involves reaching a more complex conclusion than this. It is plausible, however, that young children at a certain stage in communicative development do not go beyond such second-order meta-representations and that adults too do not systematically attend to the further meta-representational tiers involved in communication. Still, reaching this already complex attribution of an informative intention may, in many cases, provide adequate understanding, as I will now describe.

Going from the premise 'Carol says: 'It's late'' to the conclusion 'She intends me to believe that it is time to go home', is a case of inductive inference. The conclusion goes beyond the information contained in the premise in two respects: it goes from the ambiguous and incomplete linguistic sense of 'It's late' to the proposition that it is time to go home; and it goes from first-degree meta-representational attribution of an utterance to second-degree meta-representational attribution of an informative intention.

Let us look first at the meta-representational aspect. Whereas the initial premise of this inductive inference merely attributes a behaviour to the speaker, namely the production of a certain utterance, the conclusion attributes an intention to her. Computers are coming close to being able to recognise a speaker's utterance, and that is of course a remarkable achievement. However, recognising a speaker's intentions requires far more intelligence than computers have been equipped with so far.

In general, behaviours can be conceptualised as bodily movements or as realising intentions. Conceptualising voluntary behaviours as realising intentions is far more economical, more explanatory, and of greater predictive value than merely conceptualising them as bodily movements. However, in order to conceptualise behaviours in terms of underlying intentions, an organism needs the ability to entertain meta-representations. Very few animals have any meta-representational ability. There is some experimental and anecdotal evidence suggesting that chimpanzees and possibly other non-human primates do possess some such ability in a rudimentary form.

What about humans? Do they have a meta-representational ability? Do birds fly? Do fish swim? Humans can no more refrain from attributing intentions than they can from batting their eyelids. The only issue regarding humans is developmental: at what age and through what stages do meta-representational abilities develop, if they are not there from the start? There is plenty of current research on the issue, much of it assuming that the ability to attribute beliefs and intentions appears well after the development of rich verbal abilities. I should point out that this is inconsistent with the picture of verbal understanding that I am proposing here. If the picture I am presenting is correct, some ability to attribute intentions precedes the ability to communicate verbally.

How are intentions identified? In many cases, the attribution of a specific intention involves a simple pattern of inference: the behaviour of an individual is observed to have a certain desirable effect; the individual is assumed to have intended this very effect. A man shoots an arrow and kills a deer: an observer infers that it was the man's intention to kill the deer. Though often successful, this pattern of inference will sometimes fail: a behaviour may not achieve the intended effect; or a behaviour may produce a desirable effect that had not been foreseen, and therefore not been intended either. A more reliable inference pattern will see as intentional not the actual desirable effect of a behaviour but an effect that the agent may have seen as desirable and as made more probable by his behaviour. Humans easily perform inferences of this more complex kind: a man shoots an arrow and the arrow comes close to hitting a deer: all the same, an observer has no trouble inferring that it was the man's intention to kill the deer.

Capable of attributing mental states to others, humans are also capable of forming an intention to change those mental states. Peter wears the tie that Mary has given him. A knowledgeable observer recognises Peter's intention to please Mary. Bobby tiptoes to his mother and suddenly shouts 'Boo!' An observer recognises Bobby's intention to startle his mother.

Consider a behaviour intended to change the mental state of an individual, and imagine that that target individual herself observes this behaviour and recognises the underlying intention. Will this affect the chances of the intention being fulfilled? Well, it may sometime hinder, and sometime help the fulfilment of the intention. If Bobby's mother sees him coming and understand his intention to startle her, she won't be startled. On the other hand, if Mary realises that Peter intends to please her, she may be even more pleased.

Informative intentions are typically helped by being recognised by the intended audience. Communication in particular is a means of fulfilling an informative intention by making one's audience recognise it.

It is relatively easy to recognise a behaviour as intentionally informative. Verbal and other communicative behaviours are typically attention-catching, and they have to be. I cannot communicate with you if you don't pay attention. Verbal behaviour, moreover, is largely specialised for communication.

Communicative behaviour also typically calls ideas to the mind of the audience. This may be done by non-verbal as well as by verbal means. Carol might establish eye contact with John, lean her head sideways, and close her eyes for a second, in other words, she might mime falling asleep. This would bring to John's mind the idea of sleep, or of Carol sleeping. Saying to John 'It's late' would, in the same situation, bring to his mind the senses of the sentence, which all include some idea of lateness. In either scenario, John would, at this point, recognise that Carol was intending to inform him of something, but he would still have to discover what she wanted to inform him of. The ideas she has brought to his mind fall quite short of representing definite information.

We have focused so far on the meta-representational dimension of comprehension. We are now brought back to the semantic dimension: how does John, for instance, go from the ambiguous and incomplete linguistic sense of 'It's late' to the well-understood proposition that it is time to go home? I will describe three interpretation strategies that John might adopt, a naive optimistic, a cautious optimistic, and a sophisticated strategy.

NAIVE OPTIMISM

Suppose John is very trusting indeed and takes for granted that Carol is behaving both benevolently and competently. Then John can take for granted two further things: that the information Carol wants to convey to him is information worth his attention, and that the means she is using to convey that information should make it as easy as possible for him to retrieve it.

Deirdre Wilson and I have argued in our book *Relevance* that information worth one's attention is information which brings about significant cognitive effects, that is, information which, when taken together with what the individual already knows or assumes, allows inferences which would not have been possible otherwise. Such information is *relevant*, and the greater the cognitive effects it brings about, the more relevant it is. We have argued, on the other hand, that the greater the mental effort needed to acquire and process relevant information, the less relevant it is. This being so, when a hearer assumes that a speaker is benevolent and competent, he can take for granted that the information conveyed by her is relevant enough to be worth his attention, and that retrieving it should not cause unnecessary effort, for this would pointlessly diminish relevance.

In particular, John may take for granted that the linguistic senses and ideas that Carol's utterance has brought to his mind have not been evoked in vain, and that they provide an optimal starting-point for retrieving the information Carol intended to convey. Given the situation and the way in which these ideas have been evoked, one linguistic sense is more salient than the others, some ways of fleshing it out come more easily to mind, some contextual information is more easily invoked, and thus the possible interpretations are ranked in John's mind in order of ease of access. In other situations, of course, the various possible interpretations of the same utterance would be ranked differently.

All John has to do now is follow the path of least effort in constructing an interpretation of Carol's behaviour and stop when he reaches an interpretation that provides him with information relevant enough to be worth his attention. The interpretation he will reach in this way will be the one intended by Carol. Why, you may ask, should the most easily reached relevant interpretation be the true one? Why shouldn't the true interpretation be arcane and utterly boring? From a logical point of view, relevance and ease of access have nothing to do with truth. From a psychological point of view, however, the situation is different: a competent and benevolent communicator will see to it that the information she wants to convey is indeed relevant to her audience, and is more easily retrieved than other otherwise plausible interpretations So, the first accessed, relevant enough interpretation is, by these very properties, confirmed as the intended one.

On hearing Carol say 'It's late', John may not have any specific referent for 'it' that springs to mind, so the non-referring interpretation of 'it' would be favoured. The first interpretation of 'late' he can think of might be late with respect to the time at which they had promised the baby-sitter to be back home. Together with easily accessible contextual premises, for instance the knowledge that their excellent baby-sitter might not come back if promises made to her are not kept, Carol's utterance implies straightforwardly that it is time to go home. This implication makes the utterance relevant enough, and therefore John is led to conclude that Carol intends him to accept this implication, that is, to believe that it is time to go home.

The naive interpretation strategy I have just described will yield adequate comprehension whenever the speaker is indeed benevolent and, above all, competent enough to realise what is relevant and salient for her audience at the time. However, speakers are not always that competent. Suppose for instance that, unbeknownst to Carol, John has just been worrying about a delivery that should have been made that very day. The first relevant interpretation of Carol's utterance 'It's late' to come to John's mind might be that the delivery is late, and the simple inference pattern I have described would cause him to accept – wrongly – that interpretation as the intended one. Such errors of comprehension do occur. Young children in particular easily believe that one is talking about what happens to be foremost in their mind (and conversely that what they want to talk about is foremost in the minds of their listeners). Still, most of the time, these errors are avoided. This suggests that competent communicators have a more powerful interpretation strategy at their disposal.

CAUTIOUS OPTIMISM

What this more powerful interpretation strategy might be is not too hard to imagine. It is just a special case of competent attribution of intentions. A competent observer can infer that the hunter intended to kill the deer even though the arrow flew just over the animal. The observer's inference pattern consists in seeing as intentional not the actual effect of an action but an effect that the agent might have desired and expected. Similarly, a more appropriate interpretation strategy in matters of comprehension consists in attributing to the speaker an interpretation that she might have thought would be relevant enough and most easily accessed, rather than what actually happens to be the most accessible relevant interpretation. In other words, a competent hearer allows for the possibility that the speaker might have misjudged what would be most accessible and relevant to him.

I have now introduced two optimistic strategies: a naive and a more cautious one. In the naive strategy, the speaker is assumed to be benevolent and competent, and the inference pattern consists in going uncritically where the ideas suggested by the linguistic sense of the utterance will take you: look for easy relevance; assume that it was intended. The conclusion of such an inference is a second-order meta-representational attribution of a first-order meta-representational intention. However, in that inference pattern, attributions of intentions do not serve as premises. Their complex logical structure is derived but not exploited.

In the second, more cautious strategy, the speaker is assumed to be benevolent, but not necessarily competent. She may not know what is on her hearer's mind. She may therefore fail to convey relevant information, or fail to make the relevant information she intends to convey more accessible than any other possible interpretation. As in the naive strategy, the hearer should follow the path of least effort, but he should stop not at the first relevant enough interpretation that comes to mind, but at the first interpretation that the speaker might have thought would be relevant enough to him.

Suppose the interpretation that first occurs to the hearer is relevant enough to him. His next step will be to evaluate this interpretation in the light of what he knows about the speaker. Could she have expected this interpretation to occur to him? Would she have seen it as relevant enough to him? Only if the answer to both questions is yes will this interpretation be retained. Otherwise, the next accessible interpretation will be tested in the same way.

This time, second-order meta-representations may serve not just as conclusions, but also as premises. John may reject the conclusion that Carol intends to inform him that the expected delivery is late, because he believes that

> she does not know
>> that I am wondering
>>> whether the delivery took place.

He will then go to the next possible interpretation, the interpretation on which it is time to go home. This second interpretation will pass the evaluation test: yes, Carol could have expected it to occur to him and she may have found it relevant enough to be worth communicating. John, using this inference pattern, will not be misled by his own obsession with the possibly late delivery. He will understand Carol correctly.

This second, cautious pattern of inference will yield adequate comprehension whenever the speaker is benevolent, even though she might not be competent enough to realise what is relevant and salient for her audience at the time.

But now let me confide in you: speakers are not always benevolent, and communication is not always a nice affair. If you doubt me, just look at the amount of communicative energy spent on trying to convince young people that smoking will make them glamorous. Very competent, of course, but hardly benevolent. So, a truly sophisticated hearer does not assume that every communicator is benevolent. And, for that matter, a truly sophisticated communicator is not always benevolent.

Imagine the following scenario. The baby-sitter usually leaves at midnight. This time, however, Carol, thinking that the party she and John are invited to will be great fun, has asked the baby-sitter to stay until one. Carol does not enjoy the party after all, but John does. At 11.30, she says to him: 'It's late', expecting him to think that they have to go home for the baby-sitter. Carol does not know that John knows about the special arrangement she has made with the baby-sitter. Now, if John took for granted that Carol was benevolent, he would be bound to misunderstand her: she could not intend him to believe something she knows to be false and that therefore does not constitute relevant information. He would then look for another interpretation, and assume for instance that she means late with respect to some other schedule or expectation.

You might say: given what he knows, John will, in this case, drop the assumption that Carol is benevolent. True enough, but very puzzling. Remember that the two interpretation strategies I have described so far both presuppose this very assumption. If John assumes that Carol is benevolent, he will not find the right interpretation and he will have no reason to doubt Carol's benevolence. If John drops this assumption he will not find any interpretation at all. When you take benevolence for granted, you cannot recognise bad faith or lies. When you don't take benevolence for granted you cannot use optimistic interpretation strategies that presuppose benevolence. Hearers who are capable of recognising lies must be using yet another strategy.

SOPHISTICATED UNDERSTANDING

Competent hearers (who are usually also competent speakers) realise that speakers use communication to pursue their own ends, which may correspond in some respects to the ends of their audience, and differ in others.

Communication is a special way of fulfilling an informative intention. An informative intention can be fulfilled in many ways. One way is to provide evidence, genuine or spurious, of the information one wants someone to accept as true. Thus Carol might have pretended to fall asleep at the party; John might be fooled by this behaviour and come to believe that it is time to go home, without ever realising that this is exactly what Carol intended him to believe.

Another, often more practical way of fulfilling an informative intention is to provide evidence for the fact that one has that very same informative intention. Instead of pretending to fall genuinely asleep – and hurting her hosts' feelings as a side effect – Carol might, as I mentioned, mime falling asleep in a manifestly intentional way. This ostensive behaviour would give John evidence of her intention: she intends, by her behaviour, to make him realise that it is time to go home. John might well, as a result of grasping Carol's intention, end up believing that it is time to go home. After all, if someone you trust shows you that she wants you to believe something, then that gives you a reason to believe it.

Compare three ways in which Carol might try to convince John that it is time to go home: she might deceitfully pretend to fall asleep; she

might overtly mime falling asleep, or she might say 'It's late'. In terms
of outward appearances, the deceitful pretence and the overt mime
are much more similar to one another than either is to the linguistic
utterance. In terms of procedure, however, the overt mime and the
utterance fall together. They both consist in trying to fulfil Carol's
informative intention by giving John genuine evidence of that very
intention. The deceitful pretence, on the other hand, stands alone; it
consists in giving false evidence of the information Carol wants John
to accept, and no evidence whatsoever of Carol's intention.

To fulfil an informative intention by making it known is, properly
speaking, to communicate. Communication is not an accident that
happens when people have informative intentions and these informative intentions somehow become known to their intended
audience. Communication is generally intentional. An intention to
communicate, or, as we call it, a 'communicative intention', is a
second-order informative intention: the intention to make a first-
order informative intention known. A simple informative intention
is a first order meta-representation. Knowing that someone has an
informative intention is entertaining a second-order meta-representation. Having a communicative intention – that is, intending someone
to know that one has an informative intention – is entertaining a
third-order meta-representation. Attributing a communicative intention to someone is therefore entertaining a fourth-order meta-representation. Thus, if John is aware of Carol's intention, not just to cause
him to believe, but to *communicate* to him, that it is time to go home,
then he is entertaining a thought of the following complexity

> She intends
> > me to know
> > > that she intends
> > > > me to believe
> > > > > that it is time to go home.

When an agent is known to have some main intention, it can be
inferred that he also intends whatever he believes is necessary for his
main intention to be fulfilled. The hunter intending to kill the deer

intends his arrow to hit the deer in a vulnerable spot and with enough strength to penetrate it.

In the same way, when someone is known to have a communicative intention, further intentions of hers become inferable. To have a communicative intention is to hope to fulfil an informative intention by making it known to one's audience. For this to succeed, there is an obvious condition that must be fulfilled: the audience must believe that the information the communicator wants to convey is relevant enough (and is true to the extent that truth is necessary for relevance), or else the audience won't pay attention, or won't accept the information as true. It is manifest therefore, to communicator and audience alike, that the communicator intends her audience to assume that what she intends to communicate is relevant enough. In other words, every act of communication, and in particular every utterance, conveys a presumption of its own relevance. This universal fact is what we call *the principle of relevance*.

As I have argued, if a hearer can postulate that the speaker is benevolent and competent, and therefore that her utterance is relevant to him, then he has a good and simple interpretation pattern at his disposal. However the initial postulate is often false. If a hearer can postulate that the speaker is at least benevolent, and therefore that she believed that her utterance would be relevant to him, he has a good, somewhat more complex interpretation pattern at his disposal. However, this postulate may still be false. If a hearer now attributes to the speaker a fully-fledged communicative intention, then he need not *postulate*, he can *infer* that she intends her utterance to seem relevant to him; if he is right in assuming that she is communicating to him, he cannot be wrong in making this further assumption. Does this assumption provide him with a good interpretation pattern? Yes it does.

In the third, sophisticated strategy, the speaker is not assumed to be benevolent or competent. She is merely assumed to intend to seem benevolent and competent. The beauty is that now even if that intention fails, interpretation may succeed. As in the other interpretation strategies, the hearer should follow the path of least effort, but he should stop, not at the first relevant enough interpretation that comes to mind, nor at the first interpretation that the speaker might have thought would be relevant enough to him, but at the first interpretation that the speaker might have thought would *seem* relevant enough to him.

Suppose the first interpretation the hearer comes up with is not relevant to him at all, because he knows that it is false, however it would be relevant enough if it were true. The hearer will evaluate this interpretation: could the speaker have expected him to come up with it? Would she have seen it as likely to seem relevant enough to him? If the answer to both questions is yes, this interpretation will be retained. Then the hearer may well wonder: did the speaker assume that this interpretation would appear relevant to him because she herself believes it is true and relevant, or was she trying to mislead him? Either way, the speaker's *informative* intention will have failed: the hearer does not accept the intended intepretation as true. But *the communicative intention will have succeeded*: the hearer has correctly retrieved the intended interpretation.

Thus John will correctly understand that Carol intended him to believe that it was time to go home because of the baby-sitter. However he will disappoint Carol by not believing what she intended him to believe, and moreover he will correctly recognise her disingenuousness.

Fully-fledged communicative competence involves, for the speaker, being capable of having at least third-order meta-representational communicative intentions, and, for the hearer, being capable of making at least fourth-order meta-representational attributions of such communicative intentions. In fact, when irony, reported speech, and other meta-representational contents are taken into consideration, it becomes apparent that communicators juggle quite easily with still more complex meta-representations. This does not imply that communicators are conscious of the complexity of their mental representations. What it does imply is that every tier of these representations may play a role in inference. Much of everyday communication takes place between people who are benevolent to one another and who know one another well enough. In such circumstances, cautious, and even naive optimism can serve as 'default' interpretation strategies, and the higher level meta-representational tiers involved in sophisticated understanding may play no role at all. Still, when the optimistic strategies fail, a competent hearer resorts to the sophisticated strategy, and performs the complex meta-representational inferences it involves, without the slightest difficulty.

The ability to use such complex meta-representations is ignored in psychology, where the study of much simpler meta-representation is a relatively new topic. The logic involved in meta-representational inferences is also hardly ever studied. Are humans intelligent enough to know how intelligent they are?

FURTHER READING

Astington, J. Z., Harris, P. L. and Olson, D. R. (eds.), *Developing Theories of Mind*, Cambridge: Cambridge University Press 1988.

Blakemore, Diane, *Understanding Utterances*, Oxford: Blackwell 1992.

Byrne, R. W. and Whiten, A. (eds.) *Machiavellian Intelligence; Social Expertise and the Evolution of Intellect in Monkeys, Apes and Humans*, Oxford: Oxford University Press 1988.

Grice, H. P., *Studies in the Way of Words*, Cambridge, Mass.: Harvard University Press 1989.

Sperber, Dan and Wilson, Deirdre, *Relevance: Communication and Cognition*, Oxford: Blackwell 1986.

NOTES ON CONTRIBUTORS

SIMHA AROM is Head of Research in the ethnomusicology section of the research institute Oral Tradition: Languages and Civilisations (LACITO) at CNRS, Paris. Working principally in the Central African Republic, he has developed methods for understanding complex vocal and instrumental polyphonies, particularly those of the Aka Pygmies. He is the author of *African Polyphony and Polyrhythm* (Cambridge University Press 1992), and producer of thirty records, three of which have won the Grand Prix International du Disque of the Académie Charles Cros.

GEORGE BUTTERWORTH is Professor of Psychology at the University of Sussex. His research interests lie in perception, communication, and the origins of thought in infants. He is currently working on index finger pointing, a gesture species specific to humans that links speech and non-verbal communication. Among his publications are *Infancy and Epistemology* (1982), *Causes of Development* (1990), and *Context and Cognition* (1992).

DANIEL DENNETT is Distinguished Arts and Sciences Professor and Director, Center for Cognitive Studies, Tufts University. He is the author of several books on philosophy, most recently *Consciousness Explained* (1991). He is currently working on a book on the impact of Darwinian thinking in contemporary science and philosophy.

RICHARD GREGORY is Emeritus Professor of Neuropsychology at the University of Bristol, where he was Director of the Brain and Perception Laboratory in the Medical School 1970–88, and founder of the Exploratory Hands-on Science Centre. His books include *The Intelligent Eye* (1970), *The Oxford Companion to the Mind* (1987), and *Eye and Brain* (4th edition, 1990).

JEAN KHALFA, editor of the volume, is a Lecturer in Trinity College, Cambridge. Until 1992 he headed the French Cultural Delegation in Cambridge. Among his research interests are the philosophy of mind and the history of psychology.

NICHOLAS MACKINTOSH is Professor of Experimental Psychology at the University of Cambridge, and a Fellow of King's College. His research interests include the study of conditioning, discrimination learning and intelligence in animals and, as a sideline, intelligence testing in people. Among his publications are *The Psychology of Animal Learning* (1974), and *Conditioning and Associative Learning* (1983).

ROGER PENROSE is Rouse Ball Professor of Mathematics at the University of Oxford. Among his publications is *The Emperor's New Mind*, which won the 1990 Science Book Prize.

ROGER SCHANK is John Evans Professor of Electrical Engineering and Computer Science, Professor of Psychology, and Professor of Education at Northwestern University. He also directs the University's Institute for the Learning Sciences, a centre for interdisciplinary research into human learning which develops AI software solutions for the world of education. Among his fourteen books are *Dynamic Memory* (Cambridge University Press 1982) and *Tell Me a Story* (1990).

DAN SPERBER researches areas of interaction between the cognitive and social sciences at the Centre de Recherche en Épistémologie Appliquée of the École Polytechnique, Paris. In collaboration with Professor Deirdre Wilson of University College London, he has developed an influential approach to the study of human communications, relevance theory. His books include *Rethinking Symbolism* (Cambridge University Press 1975), *On Anthropological Knowledge* (Cambridge University Press 1985), and *Relevance: Communication and Cognition* (with Deirdre Wilson, 1986).

ACKNOWLEDGEMENTS

INTRODUCTION

Thanks to Janine Deschamps, Joyce Graham, and the Research Fellows and graduate students of Darwin College, for their help in organising the lectures, and to Andrew Rothwell, for his helpful comments on the text.

CHAPTER 1

All illustrations courtesy Richard Gregory. Figure 2, after C. Blakemore and P. Sutton, *Science* 166 (1969), 245–7.

CHAPTER 2

Figure 4, after J. M. Pearce, *An Introduction to Animal Cognition*, Hove: Lawrence Erlbaum 1987, 248, Fig. 7.12; Figure 5, after A. L. Cutting; Figure 6, Hugo van Lawick; Figure 7, after S. Baron-Cohen et al, *Cognition* 21 (1985), 41, Figure 1. Remaining illustrations, Nicholas Mackintosh.

CHAPTER 3

Thanks to Margaret Harris for helpful comments on the manuscript. Figure 2, after W. Schiff, *Perception: An Applied Approach*, Boston, Massachusetts: Houghton Mifflin 1980, 264, Fig. 6.25; courtesy General Electric Company, Fairfield, Connecticut; Figure 3, after E. S. Spelke, *Cognitive Psychology* 8 (1976), 553–60; Figure 4, after P. Kuhl and A. N. Meltzoff, *Science* 218 (1982), 1138–41; Figure 5, after P. J. Kellman and E. S. Spelke, *Cognitive Psychology* 15 (1983); Figure 6, after C. E. Granrud et al, *Child Development* 55 (1984), 1630–6; Figure 7, after B.I. Berthenthal et al, *Child Development* 56 (1985), 531–43; Figure 8, from N. Jarrett, 'The development of detour reaching in human infancy', unpublished MSc thesis, University of Southampton, 1987.

CHAPTER 4

The work of Roger Schank and Lawrence Birnbaum is supported in part by the Defense Advanced Research Projects Agency, monitored by the Office of Naval Research under contracts N00014–91–J–4092 and N00014–90–J–4117. The Institute for the Learning

Sciences was established in 1989 with the support of Andersen Consulting, part of the Arthur Andersen Worldwide Organization. The Institute receives additional support from Ameritech and North West Water Group plc, Institute Partners, and IBM.

CHAPTER 5

Roger Penrose's chapter is partly based on a portion of his forthcoming book *Shadows of the Mind*: Oxford University Press 1994. The models depicted in Figs. 1 and 2 are constructed using ZOMETOOL. Thanks go to Vanessa Thomas for providing the computer pictures of Figs. 3 and 4, using *Mathematica*, and to Bill McColl for information concerning automatic theorem-proving.

CHAPTER 6

Thanks to Judith Schlanger, Dany Dayan, Jean Khalfa and Serge Pahaut for helpful comments on the manuscript. Sources of quotations are as follows: Jean Molino, *Ethnolinguistique. Contributions théoriques et methodologiques*, Paris: SELAF 1981, 237–49, at 242–3; Roman Jakobson, in T. A. Sebeock (ed.), *Style and Language*, Cambridge, MA: MIT Press 1960, 350–77, at 364; Stephen A. Tyler (ed.), *Cognitive Anthropology*, Prospect Heights, IL: Waveland Press 1987, 1–23, at 6.

CHAPTER 7

Sources of quotations are as follows: p. 173, John Holland, 'Complex adaptive systems', *Daedalus*, Winter 1992, 25; p. 174, André Roche Lecours and Yves Joanette, 'Linguistic and other psychological aspects of paroxysmal aphasia', *Brain and Language* 10 (1980), 1–23.

CHAPTER 8

Dan Sperber's chapter is based on work done jointly with Deirdre Wilson. Figure 1 is from Saul Steinberg, *The New World*, New York: Harper and Row 1965.

INDEX